IN MY NATURE

Rediscover your wild and free spirit

Tanja B. Linde

Hardie Grant

EXPLORE

I am a part of nature,
and so it will always stay.
Through mind, body and spirit,
in every possible way.

It's in my nature to always grow.
It's in my nature to flow and let go.
It's in my nature to be wild and free.
It's in my nature to live happily.

All this and much more,
I've learned to be true.
In this book I'll show you,
it's in your nature, too.

Introduction

Allow me to introduce myself – I'm Tanja and I'm the author and artist behind this book. I've been on a journey with the aim of reconnecting with the nature within and around me. It's been a wild ride with magical moments, life-changing encounters, spiritual experiences and natural wonders. It all began when my partner Lenny and I decided to take a gap year, and this turned into a decade-long adventure as nomads, where we backpacked, hitchhiked, couch-surfed and camped our way through Europe and Asia. We lived a life on the go, and the only times we ever stayed put for several months were when we worked seasonal jobs to save up money for our next adventure.

My curiosity for exploring the world started when I was a kid. I was raised in three different countries, speaking three different languages and learning three different cultures all by the age of nine, so if you ask me where I am from, I can't pick if it's Ireland, Germany or Denmark. Yes, you could say I was confused growing

up, and no, I wouldn't change my childhood if I could. It's what has allowed me to know that new perspectives, new experiences and a new version of me can always be hiding around the next corner. This knowledge opened my mind to the fact that my life could head in all kinds of directions, and, oh, was I curious to see what opportunities and adventures lay ahead. This relentless curiosity is what has shaped my adult life to this point, as the open road became my home and way of life.

What started as one work and travel trip quickly turned into another, and this cycle just kept continuing year after year. With our first savings from working as waiters in Norway, we bought a 24-year-old VW van that we called Elefriend – he was big, grey and ever so slow. He became our trusted companion, and together we hit the road for three months to explore 13 countries in Central and Eastern Europe.

For the next seven years, Elefriend never let us down. Even though he had his breakdowns (don't we all though?), there was nothing a little mechanical work couldn't fix, and we were always back on the road within a day or two. Road tripping with him was how we got our first taste of freedom, and it's safe to say that we were hooked! After our first trip, we knew more adventures were to be had and more perspectives to be gained, so when we ran out of money we applied for seasonal jobs again – this time in Arosa, Switzerland. Apart from buying seasonal ski-lift tickets, we saved every penny to spend on travelling for the rest of the year. Our adventures spanned two continents far and wide – and our encounters were everywhere from thrilling to terrifying. Travelling made us rich in experiences, and we got to meet new people, learn new lessons and find new objectives anywhere we went. We couldn't get enough and always sought to experience the magical, mystical, spiritual, cultural and natural beauties of the world.

Nevertheless, jumping from one experience to the next also drained us of a lot of energy. It was not an easy-breezy life – anyone that travels on a shoestring will tell you that much – and dealing with scammers and going through unpleasant or dangerous situations on a regular basis were just some of the obstacles along the way. Nonetheless, the bad never outweighed the good, and being faced with challenges turned our journey into an all-round rich experience. As my childhood dreams of becoming an unbound nomad and exploring the world became everyday reality, travelling became my sole purpose. I got out into nature, I immersed myself in other cultures and I left my comfort zone on a daily basis.

Even though travelling made me grow and change in many ways, I kept finding myself wanting to live up to certain identities, like nomad, hippie and van-lifer, thinking staying synonymous with these labels, I would undoubtedly become my happiest self. Having placed too much value on superficial ideas, I had also become a little too convinced of the whole 'find your tribe' trend. I was too eager to change myself to fit in somewhere, and my eagerness to belong was stronger than my acceptance of myself. I couldn't just be happy on my own terms, because I didn't take the time to see what my own true nature really was. Now I can't help but laugh at the irony that I kept putting myself in boxes while attempting to become my authentic self.

What's even more ironic is that these realisations came to me when I had literally boxed myself up in a van. Lenny and I had sold dear old Elefriend to convert a Sprinter van, which we named Navi the Whale, into our own little cabin on wheels to live in full-time. Even though we enjoyed aspects of the build, it was super draining – both energy-wise and money-wise. What kept us going was the idea of what lay ahead: a life of non-stop adventure, always being free to go or stay wherever we pleased. On paper this idea answered all our hopes and dreams – in practice we came to discover that van life created more hassle and restrictions than we had ever felt before. After several exhausting months on the road, we realised that living in the van full-time was just as draining as building it! Only then did I begin to understand that the wild and free life I seemingly had didn't truly make me happy, nor free.

It took some soul searching to see that happiness and freedom start within and often aren't dependent on external factors. This realisation uncovered the fact that I had been restricting and confining myself in a number of ways. Having placed too much importance on the aesthetics of being a spiritual nomad hippie, I had left interests, dreams and plans on the side of the road that didn't fit the vision I had of my lifestyle. I had been chasing the idea of a wild and free life, not realising I had focused too much on the external aspects of it. I thought the way I dressed, the home that I lived in, the books that I read and the activities that

I did were the very things that made me a free spirit. But these material things merely created an image, a fabricated identity that I kept myself very occupied to maintain.

After having all these realisations, I made the decision that I would seek my own version of happiness and freedom. I would no longer keep myself distracted with superficial ideas and instead try to dig deeper within. I found myself heading in the direction of nature – its rhythms, vibrations, ebbs and flows were calling to me. I knew I would find answers in its elements, because out in nature I've always felt a sense of belonging – it's like I'm free to be my most natural and radiant self. I don't have to change for nature; I can just be me. Nature keeps sparking my life force time and time again, and this potent energy is also what I use to create my art. I naturally find the biggest source of inspiration to be nature, as there is endless beauty and significance to be found in the natural world.

Nature has taught me to listen to my intuition and to keep myself open to whatever comes my way. Its infinite wisdom helps me practise letting go and flowing with the motions. It connects me to my free spirit and motivates me to leave my comfort zone on a regular basis. Discovering what lies in my own nature, I've learnt not to get too attached to ideas I have about myself or what my life should be, because my track record shows I often outgrow them. I now strive to continuously grow and break free from old bonds and ideas that no longer serve me, to counteract the tendency to get too set in my (and others') ways.

Embracing transformation has become such a big part of Lenny's and my lives that we go by thebreakawayers on Instagram and live by the motto, 'Break away today'. Even though we now live in an apartment, our motto is still as true as it ever was before: it's a daily reminder to continuously break away from past versions of ourselves, from norms, from misconceptions and all the other things that take too much room within us or keep us from being our most wild and free selves.

This book is a compilation of some of the lessons and experiences I've learned throughout my journey. It's a collection of the findings, activities, rituals and practices that I apply to my daily life. They help me create inner balance, connect me with nature and help me reach new levels of my growth. You can also begin applying these starting today, without having to quit your job, buy a van or travel across the globe. Allow me to share something I've learned along the way: as long as you open yourself up to appreciate life, it doesn't matter where you are, how you're living or how much money you have; there are always new things to learn, amazing things to explore, beautiful people to meet and exciting things to try. The world offers a limitless supply of infinite wonders and they are found everywhere, even within you! This is why I hope to help you reconnect to your free spirit, as it can help you see these wonders. So without further ado, let's venture inwards. May you find what's in your nature and may you live happily and peacefully in its wholeness.

Tanja B. Linde

Venture inwards

Your wild nature is waiting to be discovered, but you don't have to search far and wide – the place to start looking is within.

I like to think of it this way: my body is a van, my mind is in the driver's seat and my spirit (see page 68) is what guides me – my inner GPS if you will. When my mind is in the right place and holds space for my spirit, I can make the right decisions on where to go and what to do. I've discovered that when I follow my spirit, which speaks to me through intuition and gut feeling, it leads me to see places I never knew existed, meet people that become my new best friends and find adventures that take me out of my comfort zone.

My mindset proved to be the thing that made all the difference, no matter which situation I found myself in. When my mind is not in the right place, it doesn't matter if I'm here, there or anywhere, because I'll not experience it fully. I cannot truly treasure moments, take chances and see opportunities if my mind is distracted with fear and confusion. But when there is balance in mind, body and spirit, I'm basically able to do what comes naturally to me, as I am living from a place of wholeness.

THE NATURE OF YOU

To understand your own mind and what makes you wild and free, you need to look into your own nature – getting back to the root of it all. For this I want you to imagine the following: your mind is like a garden where ideas, thoughts and assumptions have been planted from the day you were born. Most seeds come from your external world, your family, friends and society, yet you might tend to this garden, thinking that everything that grows here is your own.

With time, some of the seeds you cultivate will turn into beautiful flowers but others might turn into harmful weeds that go against your own wellbeing, such as ideas, beliefs and misconceptions that go against your nature.

But if you look over the hedges of your garden, you will see that there's more to you. Outside lies a wild and untamed place that holds space for peace, love and happiness. There are no borders or boundaries. No judgements, reservations or expectations. This infinite wilderness is part of your true nature and goes by many names, but is often called the spirit or higher nature (see page 68). When you make space for your spirit within, you make room for growth, transformation and expansion. As you discover this wild and free side of yourself, you can use this broadened viewpoint to differentiate between wildflowers and weeds, as your higher nature can show you your authentic self and helps you find a deeper connection to yourself and others.

Challenge your ego

Your garden is only your immediate consciousness, also called the ego. You might go through life identifying with it, thinking this mental commentary is the real and only you. Your ego is what creates sets of ideas about who you are, clings to a sense of identity and tries to find understanding of the world through constant rationalisations. It does so, thinking it is in your own best interest.

Your ego's 'personality' and its ideas, aspirations, fears, resentments, attitudes and beliefs are by and large a product of outside influence. Most of the ideas and opinions you've had throughout your life didn't actually originate within. Even the thoughts and feelings you thought defined you as a person originated from the external world. Knowing this, you can break free and turn the ego from enemy to ally. The first thing to understand is that the ego lives in your thoughts and feelings: the moment you stop identifying with your mental commentary is when you actively free yourself from believing the illusion. That is when your higher nature can turn the ego into its tool, because the ego is what mediates your conscious and unconscious self, and with the power of the ego's free will, you can align yourself to your higher self.

Your ego is attached to the physical world and always wants to meet its own needs. It has a very narrow-minded view, making it opinionated and fixated on the material. The ego is fuelled by fear, making you feel tense, contracted and alone. The voice of the ego can be heard when you feel worry, doubt, envy, guilt, anger or shame. The ego wants power and to stay in control at all times. It can have the wrong intentions, making you pursue things that feed its ego, like money, power, status and possessions. These are all things the ego thinks makes it special, and will make it get the recognition it thinks it is so deserving of.

Your higher self belongs to the non-physical world and strives towards collective needs of yourself and others. It has an expansive perspective that goes beyond the views of the ego, so it is often experienced as being neutral. The voice of your higher self is heard when your mind is quiet or it shows its presence when you feel intuition. Holding space for spirit is how you can allow wilderness into your garden, making you more wild and free. Holding space for spirit, you can cultivate your mind and body and find balance between the rational ego and the intuitive spirit.

How to invite in your higher self

- Question the nature of your ego, become an onlooker of your mind.

- Practise freeing yourself from subjective thinking, seek a higher viewpoint.

- Don't identify with your inner commentary, you are not your thoughts and feelings.

- Reclaim what you lost to your ego, rediscover your true nature.

- Whenever you feel stuck in your thoughts, shatter the ego's illusion.

- Practise meditation (see page 42) and mindfulness (see page 39).

Affirmations to silence your ego

I trust my intuition and listen to the
wisdom of my nature.
My mind is free of resistance and open
to all possibilities.
I create my own reality and live in
harmony with the world.

CHOOSE YOUR OWN PATH

Throughout the stages of our development, we learn to view everything, even ourselves, from a certain perspective. We are told what is good and bad, right and wrong, possible and impossible, responsible and irresponsible. Being susceptible is in our nature, so we take on others' opinions, lessons and beliefs and make them our own. This is one of the ways we are being indoctrinated into the ways of our world.

Tracing back the steps, most of us can find a time when we were allowed to be wild and free. When we were children, we were encouraged to be creative, imaginative and playful. We got to explore the world with childlike wonder and were free to go through phases. The process of learning, transforming and growing was a given.

With time, we were less and less encouraged to be free and were prepared for the 'fact' that reaching adulthood meant joining 'the real world'. We were told that this was the way for us to find our place in the world. Play and imagination was now swapped for responsibility and rationalism, and keeping both didn't really seem like an option – it was time to stick to a personality, pick a job and join the grind. The taming process took over and edged out all that was wild, and our priority was no longer to learn, transform, grow, create, imagine and play; it was to conform and adapt to the way of the world and the path others had laid out for us.

Through conditioning we are taught to compromise our true self and to conform to what is seen as 'normal'. We don't want to justify ourselves or be the odd one out. We are given a path, and we obediently stick to it believing it to be the most sensible way. We abide by its many rules and heed its warnings, as we believe every step outside the path will end in failure or demise. Blind and fearful we follow the path laid before us, instead of opening our eyes and questioning the direction and purpose of this path. We have become so accustomed to the fears and worries that we find comfort in staying on it, even though it might cause us to feel unhappy.

Taking a quick look at the state of the world, we are obviously heading in the wrong direction. Now I don't want to trigger fear, but the fact is that humanity on a grander scale is making choices that are leading towards demise, so, I think it's about time that we begin to question what we have been told and what we are telling ourselves. There is too much division, too much greed, too many wrong priorities, making us destroy our own chance of happiness and the chance for future generations to live in peace.

Our attempts to find direction solely by reasoning often leads us to the wrong conclusions, as our ego (see page 15) is mostly guided by fear. When we give in to our worried mind, we might pursue the wrong things, take the wrong job, stay with the wrong person and become judgemental and close-minded.

One thing that we are doing wrong is removing ourselves from nature. We are removing ourselves from each other, and we are removing ourselves from our own wilderness. We are fighting the natural way of things, because we are misguided by our own ego. But when we aim to raise our own consciousness, we can unlearn and undo the things that work against our own best interests. That's how we can be wild and free and take actions that have a more positive impact on ourselves, other people and the Earth as a whole.

How to walk your own path

- ☮ Seek your own true perception of things by questioning everything.

- ☮ Believe in yourself and understand your own values (see page 35).

- ☮ Surround yourself with people that support and encourage you.

- ☮ Only pursue what resonates with you, don't aim to please others.

- ☮ Listen to your higher nature: your intuition and inner guidance.

Born to be wild

Wild and free – that's my approach in life. It continuously offers up personal insights, limitless adventures and an ever-growing perspective of what it means to be alive, and I feel so strongly

about unleashing my free spirit that I see it as a way of life. I don't think I broke any kind of complicated code or that this approach is my idea. On the contrary, I think we are all born wild and free but are tamed along the way.

What does it really mean to be a wild and free spirit? The way I see it, being wild is the opposite of being boxed in. I would like to explain with the following example using the concept of nature: no matter how much you try, you cannot explain nature by fitting it into a box. There is no simple explanation that encompasses all of nature, so it cannot be easily understood, or boxed in, because nature is everything from the smallest molecules to the vastness of the cosmos. Being a child of nature, you mirror its complexities.

The more you open yourself up to your own wilderness, the more you can break free and be your true and wild self. Reclaiming what makes you wild might be daunting, and if you feel unnerved now, I totally get it – I was once in your shoes! But I assure you the more you do it, the more comfort you will find in exploring your wild side, as this is who you are truly meant to be. Your wild and free nature is waiting for you to be rediscovered. Now is the time that you can start to reclaim your birthright. Rise to the occasion and get back what was always yours!

Breaking free

You have been limited, confined, tamed
and restrained.
Yet your wilderness and free spirit
always remained.
You can break free from these chains,
right here and now,
By rediscovering your wild nature,
let me tell you how.

Now that I've got that rhyme out of my system, I'll shed light on some of the things that edge out your wilderness, but first a word of caution: the following notions cannot be escaped once and for all; you'll have to continuously break free from them throughout your life to counteract the tendency to get too set in your own (or others') ways.

Free yourself from labels

All the information that you perceive through communication is overwhelming, so you label and categorise just about everything in an attempt to make sense of it all.

Labels are famous for creating such misconceptions. Attempting to comprehend someone's identity by labels creates room for a lot of error. Labels are not only problematic when it comes to judging others, but a personal label can also change the view you have of yourself, and so, a single label can shape your life and change your actions, thoughts and demeanour. You begin limiting and altering yourself, either because you try to fit the description of a word or because you want to escape the label altogether.

I started to realise the power of labels when seemingly innocent words like nomad or van-lifer had taken over every aspect of my life. Wanting to live up to my idea of these labels, I always felt I was coming up short. I felt like I wasn't truly a nomad, because I knew I would like to grow roots eventually. I wasn't fully a van-lifer either, because I felt like I was cheating

when house-sitting every couple of weeks. This always left me questioning who I was. At some point, I realised my complex existence could not be boiled down into a couple of words. After all, labels cannot describe who anyone is at their core.

How to see things as they are

- ☮ Try mentally removing all labels that you use for yourself and others.

- ☮ When making judgements, catch yourself and reconsider if they are valid.

- ☮ Find new ways to describe yourself that don't alter your true nature.

- ☮ Abstain from making assumptions when you communicate with others.

- ☮ Question whether your understanding is twisted or limited.

Question everything

In order to challenge our egos (see page 15) and break free, we need to continuously question everything. This is how you open up to your wilderness and find answers in your higher nature. That is why throughout the book I'll ask you questions. They are not rhetorical, so when they arise, really try to stop and ponder how you would answer these questions. You can learn a lot about yourself from your answers, and these findings are vital, as they can help teach you how to be wild and free.

You experience all kinds of confinements throughout your life. Some are created by your ego and others are imposed on you by society.

In order to pinpoint what in particular you need to break away from, you can try to shift your focus within and try to answer the following:

☮ What confines me?

☮ What limits my being?

☮ What holds me back?

Try finding answers without judging yourself, without limiting your thoughts and without repressing your emotions. Thoughts and emotions are the very places you can begin the search for your answers.

The answers you think of might be scattered on a wide spectrum. In some ways your confinements are simply the areas where you have a responsibility or an obligation. This is the case when it comes to your commitments, like your job, relationships, schedule and bank balance, which all bind you in some way or another, but these are not necessarily blocking you from living on your wild side.

Contemplate what stands in the way of your ability to be happy and at peace. When you follow the breadcrumb trail of what holds you back, you can discover what truly confines you and causes your free spirit to be trapped. It's a difficult and confusing process, as your thoughts can easily misguide you, and you might not find clear answers right away, but keep the questions open and the answers might begin to unfold within the next few days or weeks.

Try to figure out how you would end the following sentences. There are no wrong answers, but make sure to ponder on each to find an answer you feel comes closest to the truth.

- ☮ I'm worried about ...

- ☮ I'm restricted when ...

- ☮ I have a limited ability to ...

- ☮ I hold big expectations when ...

- ☮ I feel trapped because ...

- ☮ I keep thinking about ...

- ☮ I can't be myself when ...

By answering these questions, you might have noticed that you struggle to pinpoint what your best answer is. Either you can't decide between answers or your higher self and ego create an inner struggle. The way you feel at any given moment deeply affects your current perception of reality and changes how you respond.

I started noticing this when one day I would hold one belief about something and the next it was the complete opposite. That's when I realised that whenever I felt stressed, frustrated and lost, I would believe one thing to be true, but when I was carefree and at ease, I would believe another. You can learn to understand how you respond by checking your mood and looking at the nature of your answers.

Seek your higher perspective

As you become familiar with identifying your thought processes, you can learn to view everything from an expanded viewpoint, instead of narrowly looking at whatever crosses your mind. This allows you to watch, guide and change the direction of your mind. Through this process, you become aligned with your free spirit, enabling you to make more sensible choices from this expanded and elevated perspective.

We are constantly changing, evolving and growing. In one moment, something can be our source of motivation and inspiration, and in the next, it can turn into the reason we want to head in a completely different direction. Just because we believe something is the right thing to do in one moment doesn't mean it actually stays this way. And that is okay! You are not bound by the ideas of your ego (see page 15); you are not supposed to stay the same. Your thoughts, ideas, opinions, beliefs and life approach can change throughout life. This openness and the willingness to adapt is what allows you to be wild.

When you listen to your higher nature, you can refuse to give in to your ego's reluctance to change. You have to go with the motions, adapt to circumstances and make the most of your journey in life. You can do so by making the right kind of choices, guided by your higher nature. Your free spirit has your best interest at heart and connects you to everything above and beyond yourself.

NAVIGATING THROUGH LIFE

Every day we are faced with a new string of choices and decisions – some are resolved instinctually while others need to be made more consciously. On every dawning day, hundreds of choices can be made, leading you to take action, make a change or take a chance. You can venture out into the unknown, take a leap of faith and let go of control, or you can choose to do nothing, stay put and just focus on being. All these actions happen as a result of you opting for one option or another, and so, you either stay or go, take action or relax, take control or let go.

There is only one moment that you can navigate through life and that is the moment of now. This is the only time you have the power to make choices, and ideally these should be influenced by your ability to believe, trust and follow your inner guidance, because your ego's (see page 15) attempts at reasoning might very well go against the direction that your higher nature is pointing you towards.

With a little practice, I've learned that my higher nature can serve as a compass when navigating through life. Trusting it has allowed me to head in directions that turned out to be in my own best interest. When I was on the open road or out in nature, reason only got me so far. Often my gut feeling helped me make the right kind of choices, that turned out to be best either for my own safety or brought me enriching experiences.

The times where I didn't listen to my gut and decided plainly based on my ego's attempts at reasoning, afterwards, I often felt I hadn't made the right choice. That's how I tore my meniscus while trying to push our van out of the mud, even though I had this strong feeling that something would go wrong. These are the times I want to kick myself and keep thinking, 'you knew there was something off, you should have listened to your higher self!'

I believe we can make the best kind of choices from an expansive perspective that holds room for both intuition and reason. Coming up are guiding tools that are the centre of my decision-making process.

Your centre

At the centre of your existence is your reality. It's solely found in the present moment. Using your mind's eye to look back at your past, you can only see the projection your mind has created, which is influenced by everything you have ever felt, thought and experienced leading up to this point in time. Every time you try to predict the future, all you do is create new imaginary projections that might never eventuate. All these projections are moving your mind away from living in your present wholeness. That's why you should aim to keep both feet planted in your current reality, as it's at the centre of your existence. Only from here on out can you take actions and make choices on what to do next.

Affirmations to stay centred

☮ I am rooted in my current reality.

☮ I am living in my wholeness.

☮ I stay at the centre of my existence.

Your values

Values take form when you believe something holds more worth than something else. Values help you decide how you want to spend your time, who you want to spend it with and what your goals are. Through your free choice you need to make decisions on a regular basis, and your values factor into that equation, allowing you to choose between this or that.

Your values are the base of many decisions. Yet your values can be misguided by your ego; you might value money more than people, possessions more than memories and recognition more than achievement. The ego wants to be praised, acknowledged and taken notice of, as it wants to be worthy of other's attention. What you believe to have more or less worth is heavily influenced by your ego, so it's a good idea to study and question your values to see if they truly align with your higher nature.

Question your values

Contemplate your values and write them down. Then identify central themes, and use these to create five core values that you hold.

Your instincts

It's in your nature to act and react in certain ways. You were born with a set of instincts, which is part of your behavioural pattern. One of the most famous is the fight-or-flight instinct that happens when your body registers danger, real or perceived, and it's an evolutionary response that has kept our ancestors alive in the wild. This instinct is natural and important for our 'wild selves', but the constant stressors of modern society are tricking us into this response.

The fight-or-flight response kicks in when your wandering mind tries to predict the future in an attempt to be prepared for any imaginable obstacle, and when you live in a state of uncertainty, stress or fear. Your ego (see page 15) is always looking for potential problems, dangers and threats. You may think this is in your own best interest, not realising that in doing so, you are limiting your beliefs and fuelling your fears.

When you give in to your fears, your judgement becomes clouded by your worries. When you're not occupied by worries, you can move beyond your fears and place trust in your higher self (see page 15). By choosing to believe that everything is going to be okay, you can stay centred in the here and now, and from this calm and collected place, you can take better action and make more sensible decisions.

Five steps to let go of fear

1 Acknowledge what fears and worries occupy you.

2 Contemplate if and how they are grounded in reality.

3 Notice in which ways your mind predicts worst-case scenarios.

4 Now imagine the best-case scenario, which dissolves all your fears.

5 Choose to believe in the best-case scenario.

When you practise this process, you can begin catching your worries early on, thus changing the course of your wandering mind.

Your intuition

Intuition is one of the ways your higher self communicates with you, and it does so through your feelings. It might be a gut feeling, a hunch or a pull. When you listen to your intuition, you can find reassurance beyond the findings of your ego's reasonings. Knowing your values will help you make rational decisions in everyday life, but you cannot only rely on what seems logical, as you wouldn't be listening to what feels right.

But intuition does not guide you on a path that is all sunshine and rainbows; it might lead you to face the things you need to overcome or change in order to grow, so following it might bring forth a revelation helping you see your life in a new light.

The best choices are taken from a place that holds space for both reason and intuition. To actually get to this place, you need to try to silence your ego (see page 15), as it creates an inner noise that makes you incapable of just feeling. Its logical attempts to base choices on pros and cons won't let your intuition have a chance to speak. When you practise ways to still the mind (see page 42), you will hear the voice of your higher self.

Activity: guidance from your higher self

Your inner compass helps you navigate everyday situations and guides you in times of complete chaos. Write the following down on a piece of paper and carry it with you as a reminder.

- ☮ I stay present with my truth
- ☮ I value myself and my time
- ☮ I choose faith above fear
- ☮ I listen to my intuition

LIVE MINDFULLY

Your wandering mind can head in all kinds of directions, and if you are not careful, it can get stuck reliving the past or conjure up worrisome visions about the future. It can also lead you to envious, monotonous and hurtful thought patterns making you feel depressed and unmotivated. This is how your thoughts can change the way you experience reality.

Every time your mind wanders off you are not living in your wholeness – you have to get out of your head and into this present moment. Arriving in the here and now, you can become mindful of everything and everyone that surrounds you. This is basically the idea of a practice called mindfulness, which aims to bring your attention and awareness to the present moment and stay there.

I first learned about this practice back in 2014 when I did my first retreat in Thailand. I was very nervous going into this experience, as it was completely new territory for me. In fact, I had never heard of the practice of mindfulness before arriving there. The yoga-meditation program they offered was designed to slowly lean into the practices of mindfulness. We would spend the mornings in complete silence and do yoga to prepare for meditation. After sitting in meditation, we cooked and ate our food while staying in mindful silence. After that we were allowed to talk, but were encouraged to use the teachings of mindfulness for the duration of the day. It was a great opportunity for me to see that I had become so used to constantly being occupied with my mind that stilling it was nearly impossible. I realised my mind was an endless mental commentary, constantly thinking, reflecting, planning and dreaming, all the while I was forgetting to be in the here and now.

Practising mindfulness is something I sometimes still struggle with to this day, but this practice has significantly improved my mental wellbeing.

Ways to practise mindfulness

You can try mindfulness right now by doing a check: use your senses to notice five things in your current reality. What do you see, hear, taste, feel and smell? Try implementing mindfulness into your daily life. The more you experience your current reality via your senses, the more you stay aware and conscious of the here and now.

Five ways to practise mindfulness

- ☮ Bring attention to your breathing.

- ☮ Close your eyes and just listen to your surroundings.

- ☮ Eat slowly and savour every bite.

- ☮ Stretch your body and observe the sensations.

- ☮ Become aware of your thoughts and feelings.

Your thoughts: still your mind

Meditation practices allow you to take breaks from your thoughts or can help you gain back control of your mind, instead of having it control you. By implementing methods of meditation, you can practise to stop the wandering of your mind. Many spiritual practices agree that this should be the sole motivation behind meditation, but there are some who think that meditation is a means to reach enlightenment. By making enlightenment a goal of meditation, the intention is no longer just to still the mind.

According to Buddhist meditation practice zazen, the aim should be to have no goal when sitting down. The Zen masters teach that the practice of just sitting is just that – sitting. There is no other reward and no other intention than just sitting without thoughts.

Ways to quiet your mind

Meditation can be as simple as just sitting, yet many never get to the point of implementing it into their routine, even when they know it might benefit their wellbeing. I have a theory for why this is, and it's based on my personal experiences. Going from a busy mind to just sitting down and being still is a big leap. I can get so accustomed to always doing something, thinking about something or being entertained by something. Doing nothing feels uncomfortable and drastic, as it's the complete opposite of doing everything.

I have learned that I can take smaller steps to quiet my mind, which can pave the way to doing nothing. Meditative states don't only occur when sitting cross-legged with your eyes closed; they can happen naturally through many different kinds of activities. In my personal experience, I often enter a meditative state while painting and drawing. My mind is not completely still, but it's actively resting and not busy with thoughts. Running, gardening, knitting, stitching, playing an instrument, wood-working, cooking and colouring are other activities that can induce active resting.

Another way to achieve a quiet mind is via meditative chanting, which is a practice where certain words and syllables are repeated. Listening and taking part in repeating chants can help you stay present in the moment.

Chanting meditation

- ☮ Sit or lie down so that you are comfortable.

- ☮ Either keep your eyes open or closed.

- ☮ Listen to a simple chant (see tips below).

- ☮ Stay with the words and calm your breath.

- ☮ Don't let your mind wander away.

- ☮ If your mind starts to wander, bring your awareness back to the present moment.

- ☮ Join in the chant when you feel ready.

- ☮ Align your breathing with the chanting cycle.

- ☮ Stay in this cycle as long as you like.

Three universal chants

1 *Om* – serves to open and clear the mind.

2 *I am* – honours yourself and who you are.

3 *So hum* – unifies you with the universe, as it guides and protects, and means 'I am one with all'.

Chanting cycles can be found on YouTube and Spotify. Choose a chant that has an intention that resonates with you in the present moment.

INTROSPECTION: UNEARTH YOUR NATURE

By practising the different techniques from the past chapters, like mindfulness (see page 39) and meditation (see page 42), you will already have begun the process that unearths your nature, but before we proceed, I want to remind you of something: love, peace and happiness all lie within your nature, but you wouldn't know what they were if you didn't have an idea of hate, conflict and depression – being alive means you can experience everything from turmoil to bliss. I say this now, as I want you to embrace everything coming your way or anything that has crossed your path; they are part of the journey that uncovers your nature.

The following questions will help you unearth your true nature. Even if you don't know the answer to all these questions at this very moment, you will gain more clarity about yourself by watching and analysing your thought process. Ponder on each question and try to dig deep. While answering these questions, you might be worried that you will unearth something dark and uncomfortable. Even so, I encourage you to answer them as they will push you out of your comfort zone. It's noteworthy that you cannot dig for any specific results, you have to keep an open mind and see where the process takes you. In this way you can always become the best version of yourself, without the ego getting in the way.

Unearthing questions

What truly makes me happy?
What are my deepest wants and needs?
What makes me feel fulfilled?
Am I doing what I love?
In which ways am I limiting myself?
What do I think of myself?
What do others think of me?
Which fears and worries
occupy me?
What resentments do I carry?
Do I have unresolved issues
or traumas?

As you evaluate your inner findings, is there something you feel that can be resolved? Are your findings giving you a sense of unfinished business or unresolved traumas? Do you need to forgive yourself or someone else? Are there any answers that go against your general intention in life? Do you feel that a certain behaviour is holding you back from pursuing the things that you are passionate about? Are there ways you're self-sabotaging that stand in the way of your happiness?

Illuminate your inner world

Since you are always evolving, the act of unearthing and asking questions is not a one-time thing. It's to be practised continuously throughout your life, so that you can illuminate your inner world as you grow and evolve. This process is all about gaining clarity and raising your inner awareness. You will not resolve everything all at once, as most things take time and work to be dealt with, but by asking the questions, you open up a dialogue with yourself and your higher nature, which shall remain an open conversation. Because the moment you think you know everything, you may no longer hear what calls for your attention. You don't notice warnings or see signals. Always aim to keep open and aware of your internal processes.

When we want to ignore certain things in our life, we often go through the day just telling ourselves 'Everything is fine' and 'I shouldn't complain'. We bury our thoughts and emotions, hoping they will decompose and disappear, but they have a way of transforming into weeds of resentment that begin growing in places where they have no place. We might as well just deal with the matter at hand sooner rather than later, instead of creating a burial ground for the things we deem uncomfortable or unnecessary.

I've usually justified the burial of my thoughts and emotions by thinking that other people are living worse realities. Comparing myself to others, I said to myself to just suck it up and live with whatever bothered me. Ironically, when it came to others, I was open to talk about their issues and thought they were always valid. I would try to help them, share my advice and try to be there for them. I didn't think or say, 'Oh, just suck it up, you are totally fine, don't think about it.'

So how could I justify saying this to myself if I wouldn't say it to a loved one? At a point, I realised it was a lack of self-love. I neglected myself, as I didn't place enough value on my own wellbeing. Through the process of unearthing, I learned to love myself, and it turned out to be the most important form of self-preservation. Only with self-love did I act in ways that were in my own best interest and looked out for my own needs. My ability to love myself creates vitality from the inside out. When my body, mind and spirit are healthy and flourishing, I can be a better support for others, as I'm stable and can share my abundance.

Choose what inhabits you

Within your nature you might have creatures in the form of bad habits that you would like to rid yourself of. These unwanted inhabitants occupy your being and keep crawling in, no matter how much you want them to leave your grounds.

Imagination can be a powerful tool when it comes to changing your behavioural pattern. Whenever you detect a bad habit, start the following imaginative process, so you begin to take action against what is damaging to yourself or others. The first step is to acknowledge your bad habit and become aware of when and how it makes an appearance. Instead of being angry or frustrated with its existence, simply view the bad habit as an unwelcome critter.

This method gives a persona to your impulse; you are no longer only talking about a concept but can name and envision the culprit. Knowing its name, appearance and offences, you can start to imagine how to exterminate the bad habit, or in kinder terms, ask it to leave and say it's no longer welcome within you. Now bring to mind how your life would improve after this critter has gone. Feel the relief as intensely as you can.

Validate yourself

I have a habit of running away from my problems. This only became evident to me at a time when I was burning out from overworking and feeling completely drained. Arriving at this

low point in my life, I realised I still had unresolved issues rooted in traumas from my past. For many years I had told myself, 'I'm fine. I'm over what happened, and I'm not really bothered by my current problems. I can deal with them, and I've lived through much worse times.' Yeah, I had a whole inner dialogue trying to reassure myself that I was fine when I actually wasn't, and this translated into the choices I made when it came to living my life.

Whenever I would go 'home', all the family troubles from the past and present became my everyday reality, so hitting the road or backpacking in far-off lands became my sweet escape. Instead of facing the problems head on, I chose not to linger around and head off for new adventures. Unearthing this tendency about my nature made me face my own traumas, resolve problems and end toxic relationships.

Focusing on resolving what torments you can release you from your own self-neglect. It's important to focus on what needs healing and give it the attention and time it needs. And I'll just say this once again to make it clear: the moment you have to tell yourself you are fine, when you know that you are not, is a clear sign that you should stop and validate yourself. Don't look away; tend to your needs and be there for yourself.

It's vital to validate yourself and bring what troubles you deep within to surface. Identifying the root is just the beginning, thereafter comes the actual work that can make you overcome, recover and heal from the past. Through introspection and professional help, you can work towards finding a healthy inner balance, which is a process that can't be rushed.

For other methods to help you unearth your nature even further, you can look into shadow work, approach your process through a darkness meditation retreat or maybe you are open to the idea of uncovering your past lives with a trained hypnotherapist. You can also try to learn more about the nature of your ego, through the personality spectrum called the enneagram. In this day and age, there are tons of ways to unearth and illuminate one's nature, so do some research or talk with your therapist about methods you can use to heal from the past or help you cope with the present.

If you are thinking that the hassle of digging up what you buried is not worth your time, let me tell you this: resolving issues, healing from traumas and overcoming struggles only makes you grow stronger. And when you can endure more, you are less likely to be wavered by the little things, as you're more rooted in a solid ground. As they say: rough winds grow tough trees.

What validating yourself can sound like

I have things that I'm going through and it's tough.

I'm feeling overwhelmed and could use some help.

I trust myself and inner guidance, even if
I sometimes make mistakes.

I'm not feeling okay, and that is okay.

I'm doing my best and that is enough.

I'm not liked by everyone but that is okay.

I'm worthy of my own love. I deserve happiness.

I'm allowed to be different, even if people
don't understand me.

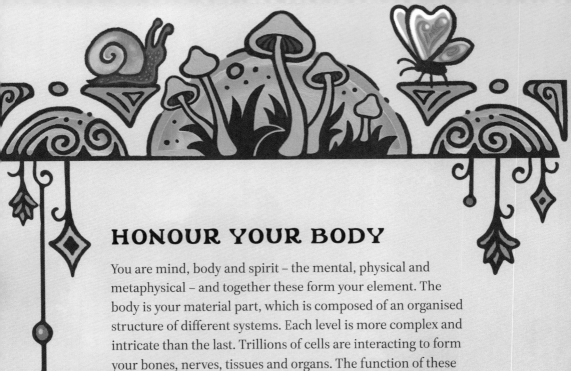

HONOUR YOUR BODY

You are mind, body and spirit – the mental, physical and metaphysical – and together these form your element. The body is your material part, which is composed of an organised structure of different systems. Each level is more complex and intricate than the last. Trillions of cells are interacting to form your bones, nerves, tissues and organs. The function of these systems and your bodily health is influenced by many factors: what you eat, how you sleep and how much you exercise all play important roles, and so do your genetics, habits and the environments you are exposed to.

Often, we don't place enough value on the function of our bodies and mainly place value on physical appearance. Impossible beauty standards make us dissatisfied with how we look, and health problems can make us want to trade bodies if we could. Our idea of what a perfect body is might disconnect us from our own just because it is not how we want it to be. We shall not feel trapped within our own flesh and matter. Our bodies are our vehicles of life, without them we would not live and breathe – we would simply not exist. We need to be appreciative of the matter we are made of and we can do so by cultivating it from the inside out.

Ways to cultivate your body

Drink plenty of water: Yes, you have heard this before, but remember it! Water replenishes and regenerates you from within. When you are hydrated, you balance out your temperature and bodily fluids. It is essential to your wellbeing to drink plenty of water, as this cultivates your mind and body, making you think more clearly and maintaining your health.

Eat the rainbow: Don't live to eat, eat to live. Food is your source of energy, so make sure to eat right. A colourful diet will give you nutrients to support your health, as each colour found in fruits and vegetables has a set of health benefits. Adopt a nutritious diet centred around a variety of plant foods. By consuming plant-based wholefoods, you can reduce the chance of evolving long-term health problems.

Practise embodiment: Reconnect with your body. Stretch yourself from top to toe and do a body scan (see page 65). Become aware of your body's signals, acknowledge any aches and pains and attend to these. Practise embodiment exercises to develop a better connection to your physical body. Exercise on a regular basis – the World Health Organisation recommends at least 30 minutes of movement every day.

Reduce toxins: We are living in a time where we expose ourselves to a cocktail of toxins on a daily basis. Pollution, bad water, chemical products (everything from parabens in beauty products to artificially fragrant air-fresheners), and microplastics are just some of many harmful things we are exposed to. Start thinking about what you are regularly exposing yourself to. Do some research to find out what might cause harm, figure out alternatives and make the necessary changes. Reduce as many toxins as possible.

BE IN YOUR ELEMENT

In many ancient spiritual traditions and beliefs, five elements – fire, water, earth, air and spirit – are said to be the fundamental building blocks that make up all matter in our universe, and they are responsible for the never-ending cycle of creation and destruction of everything in existence.

Each of the elements are distinct spiritual energies that can be invoked to deepen the connection with ourselves and everything around us, and we can use them to transform our mental health, our physical wellbeing and our way of living.

I learned about the philosophy of the five elements on my last trip to the UK, where I had the pleasure of drinking tea with a nomadic druid who was my neighbour while camping out at Stonehenge for the spring equinox. It took a little while for me to muster up the courage to knock on his door, as he had the most mystical looking van and I had no idea who was sitting inside. When I finally did knock, I was warmly welcomed inside the van, which, as I entered, had me in awe. The interior was just as beautifully crafted and decorated as the outside. There was no electricity or modern amenities, and the flickering candles made the mood even more mystic. The druid nomad, whose name was John, was an elderly man that roamed around with only the essentials, getting by on donations for his druid work and wise guidance. Speaking with John sparked my interest in the pagan philosophies of the elements, as they centre around nature.

59

Viewing our life through the five elements is a simplistic approach to understanding complex natural systems. By inspecting them individually, you can learn to use the elements to transform yourself and create balance and harmony in all aspects of your life. The elements teach us that we have a choice to either open ourselves to transform, expand, evolve, grow and transcend, or we can stay set in our ways, be reluctant to change, remain static and isolated and disconnected from the natural process that is change.

But the elements can also get out of balance. For example, when you're too wavered by your emotions, you are overpowered by the element of water, and if you are out of touch with your own body, you are disconnected from earth.

Each individual element can be a means to raise inner awareness by taking note of our mental, physical and spiritual state. We can do so by doing a self-check and seeing what occupies our mind and the emotions we feel, as well as paying attention to what sensations and signals the body is sending us. This can help us learn how our inner harmony can be restored again and again.

Fire

Fire represents change, inspiration, strength, love, sexuality, energy, life force, renewal, self-healing, stress, protection, destruction of negativity, purification and relationships with yourself and others. The element of fire is what ignites you – it's the spark of life.

Its nature is transformational, as it's the element that initiates change and gives you a potent life force in the form of passion, creativity, dedication, motivation and courage. Fire is also what makes you burn with love, which is an extremely raw and powerful energy.

You can connect to the element of fire by lighting a bonfire or a candle and gazing into the flame. Research shows that looking into the flame of a fire can have a calming effect, promote relaxation and lower blood pressure. You can also connect to this element by spending time in the sun or by intentionally raising the heat of your body with exercise and movement.

Fire is also the element of self-love, which is what makes you content in your mind and comfortable in your body. Your ability to hold and share love will enrich both you and everyone around you. Self-love in its pure state isn't self-indulgent; it's the most important form of self-preservation and allows you to be your true glowing self. Even in companionship, friendship, family and community, you need to burn for yourself and tend to your own fire before you can be a light for others. Self-love might take some practice, but you are worth it!

Activity: Write a love letter to yourself

This process will help you trigger positive emotions towards yourself, and help you become less judgemental of yourself. By proclaiming self-love, you validate yourself, and this allows you to be strong and energised to take positive action, including pursuing a passion project or living with purpose

(see page 153). Address the letter to 'Me, Myself and I'. Write down what you love about yourself and why. Add what makes you thankful for being you. Lastly, add a P.S. section for all the additional thoughts you want to jot down later on. Store the letter in a place where others won't find it, so that you write as freely and honestly as possible. Every time you have self-doubt, feel discouraged or angry with yourself, read this letter to rekindle feelings of self-love.

Water

Water represents feelings, emotions, intuition, purification, wisdom, healing, psychic ability, eternal movement, the subconscious mind, cleansing, sorrow, reflection, fertility, friendship, happiness and dreams.

Water is also a life-sustaining element – it cleanses, regenerates and restores you. You can connect physically to the element of water by going swimming. Natural bodies of water, such as the ocean, lakes and rivers, are best, but swimming pools, baths and showers work as well. Notice how you feel before you dive in, during your swim and after. You can also take a walk in the rain – or dance! – or listen to the sounds of it falling outside.

Water is always in motion, making it the element of emotions, receptivity and healing. The nature of your feelings shape how you think, act and respond in situations, and mindfulness is a great way to become observant of the flow of your emotions.

Feelings appear on a wide spectrum through many variations and endless combinations.

When you merely observe your feelings, instead of identifying with them, you can learn to stabilise yourself. The more you understand how your feelings come and go, the more you can understand the nature of your mind. You are not your feelings; you only experience them. By learning to observe them, without getting thrown by them, you can face them for what they are. Knowing your feelings, you can more easily figure out if you have any unmet needs. Acknowledge the bad feelings and never disregard them, they are just as valid as the good ones!

Gratitude routine

When you go to sleep, think of your three favourite moments of the day, as this triggers positive emotions. You can make a habit of writing them down or exchange them with a loved one to share the positive energy. Then list five things you are grateful for. You can use the following sentences as a way to bring forth your appreciative thoughts:

I appreciate ...
I'm grateful for ...
I'm thankful for ...
I love ...
I'm lucky because ...

Earth

Earth represents stability, strength, fertility, security, permanence, prosperity, wisdom, practicality, responsibility, wealth, patience, materialism, abundance and truth.

Earth is what keeps you grounded and present in the here and now. It's the foundation of life and what supports you and gives you stability, and the connection to your life path and family roots. Its material form is found in all things physical. Earth is where you come from and where you return to. By being connected to earth you are rooted in reality and present in the moment.

When you are out of touch with this element you might find it difficult to find appreciation for your own being, your current circumstances or your life in general. It's in your ego's nature to look for things that you lack, so you might constantly compare yourself with others. But when you overthink what you might be missing in your life you don't pay enough attention to what is already there.

The best way to connect to the element of earth and experience its grounding and healing properties is to physically connect to it to allow the energies to flow through you. Go outside, take off your shoes and walk barefoot on the ground. You can also lie down on the grass with your palms flat on the ground or touch – or even hug – a tree. Hiking is also a great way to experience what this element has to offer, but you can simply spend time in your garden or a nearby park as well.

Grounding exercise in nature

1 Sit down on the ground and close your eyes.

2 Inhale and exhale five times.

3 Notice the sensations you feel in your body.

4 Observe your thoughts and emotions.

5 Let your thoughts go and become present in this moment.

6 Still your mind and just take in the sensations.

7 Sit like this for a while and then rub your hands together.

8 Place your hands over your eyes and feel the warmth.

9 Take your hands away and slowly open your eyes.

10 Sit and watch nature for a while.

Do a body scan

Take ten deep breaths to relax and then shift your focus to the different parts of your body. Become aware of the sensations and simply take note of them. Start from the top and move down towards your toes. Then reverse the direction of your scan from toe to top.

Air

Air represents the mind, intelligence, inspiration, communication, imagination, ideas, knowledge, dreams and wishes, intuition, freedom, travel, psychic powers, telepathy, mental intention, finding lost things and new beginnings.

Your thoughts are the language of your mind, and you need to watch, listen and question this mental commentary. When your mind is in a good place, your thoughts can elevate your mood, especially when you feel worried or depressed. Positive thoughts can keep you calm, but your thoughts can also become stagnant, turning into bad air that can suffocate your thoughts. When you air out your thoughts, either to a trusted person or through a breathing exercise, you vent some of the bad air that takes up room within you.

You can connect to the element of air by practising breathing exercises, like those found in meditation and yoga practices, or you can simply go outside and feel the wind on your skin or watch the clouds move across the sky or the trees swaying in the wind.

Breathe out the bad, breathe in the good

The following breathing exercise is especially effective when you are feeling overwhelmed by thoughts that drag you down, bringing in more positive thoughts and helping you to be in a better state of mind to resolve and overcome the obstacles that are causing inner turbulence.

1   Close your eyes and calm your breathing.

2   Watch and identify your thoughts: are they based in stress, worry, or sadness? How are they making you feel?

3   Now every time you exhale, visualise how you are breathing out these thoughts. Breathe out the negativity to release it from your body.

4   With every breath you take, visualise positive thoughts. What are the things that uplift you, calm you and make you feel good? Visualise the air to be vibrant and glowing, and as you inhale, absorb this energy and see yourself as you become vibrant and glowing.

Purify the air

The emotional residue phenomenon is the belief that unseen emotion can linger in the air. Research shows that our sweat glands emit different chemicals based on what emotion we have, and the phenomenon states that traces of emotion might be contained in the air and picked up on and sensed at a later time. When 'bad' emotions like stress, anger and sadness happen, they might create 'bad air', which can later affect and alter your mood, so be sure to air out your home regularly. To further oxidise the air and improve air quality, keep houseplants in every room. Some of the highest oxygen producing indoor plants are areca palms, aloe vera, snake plants, spider plants and peace lilies.

Spirit

Spirit, also known as quintessence and ether, represents transcendence, intuition, energy, transformation, oneness, a sense of joy and union, and a collective consciousness.

Everything is vibrating with a universal infinite energy, and this idea is shared by many ancient philosophies and religions around the world. Chinese Taoism calls it qi, Indian philosophies call it chakra and Christians might name this energy the soul. According to pagan spiritual beliefs, this energy is often referred to as spirit, and it's the first energy that created everything in the universe. The spirit is unbound, timeless and eternal. It's universally connecting you to everything within and beyond, above and below. The spirit is what unifies you with all there is and is the one and only source of all vibrating energies. The spirit is love, truth and unity. It's you, me and everything. It's all that has been and ever will be.

Spirit is the only non-physical element and is at the centre and all around, within and beyond us. Spirit represents a higher and collective consciousness and creates a bridge between the physical and the spiritual. The element of spirit connects you to yourself, to others and to the world. It allows you to break free from separatist thinking and enables you to gain a broader and more inclusive viewpoint.

The spirit's existence creates beautiful wonders in the form of love, peace, and happiness. It's what essentially drives you to make good choices for yourself and the world and gifts you with hope and purpose. The spirit unifies all the elements you are made of and is therefore elemental to your existence. It's your authentic self and your higher nature. Unbound by time and space, it's the essence of your life, a pure energy that has no beginning and no end. It's your life force that connects you to all that is and ever was, past, present and future, seen and unseen. It also appears when you are feeling joy and peace, giving name to the notion 'to be in good spirits'.

Connect to your spirit

Connecting to the element of spirit can be difficult in our busy modern world, but can be achieved through meditation (see page 42), prayer and other spiritual practices. You can also practise mindfulness (see page 39) or engage in activities that are intuitive and creative by nature (see page 97). Have a dedicated space to co-create with the spirit and bring the elemental forces together. It can be a little space on a shelf or a dedicated shrine where you place your intentions, voice your gratitude or pray. This space is about opening the dialogue with the wilderness within and around you, making it the place for healing, transformation and manifestation.

Embodying the five elements

☮ Earth: Take off your shoes and stand up and notice how you feel steady and supported. Think to yourself: 'I'm grounded.'

☮ Air: Take three deep breaths and feel how your thoughts calm down. Visualise how each breath of fresh air is reviving your entire body.

☮ Water: Drink a glass of water and imagine the water flowing within you, replenishing all levels of your being, revitalising your life force.

☮ Fire: Light a candle and take in the light and warmth it emanates. Feel how your body takes in this energy and radiates it outwards.

☮ Spirit: Thank your higher nature and count your blessings. See yourself as you are at this very moment – safe, present and grounded.

Go explore

Exciting opportunities in all kinds of variations are right at your doorstep. You just have to cross the threshold and take a step outside to see what wonders there are to behold. Now is the time we begin following the road outwards, in the direction of exploration, adventure and opportunities. Nature is calling, so you better go!

Are you ready to step outside and explore? Are you ready to experience all wilderness has to offer? Are you ready to take the opportunities that cross your path or, better yet, make your own? If you answered yes to all of these, great! If not, even better! This will help me debunk a common myth: 'Ready or not' is an idea your brain keeps tricking you into believing as you try to predict anything that might be coming your way. The truth is, you can never really be ready for anything, and that is okay. Knowing this, you can fight the urge to obsessively prepare and take the leap into the unknown.

In my experience, enthusiasm will take you further than making a ton of plans. Instead of keeping busy with to-do lists or itineraries, you get to be open to whatever comes your way and choose to do whatever feels right. Exploring and travelling are activities that can otherwise quickly get exhausting, as you need to deal with new impressions in the here and now. So, by making no real plans, you free up mental space, which reduces stress, meaning you obsess less about the outcome of your exploration. As long as you are energised and optimistic, you have enough fuel to get you to interesting experiences and exciting adventures.

Remember to bring all the lessons from the first part of this book along when you go out exploring, as this will help you make the most of what is to come.

FUEL UP ON INSPIRATION

Ramp yourself up to explore by fuelling yourself with inspiration. Being able to visualise an action, mood or setting can be very powerful, as having a vision can set your plans and ideas into motion, which is necessary to get where you want to go. Seek out what inspires you on social platforms and in TV shows, movies, podcasts, magazines and books.

Oh, and of course music is also a great way to ramp up motivation! Which is why I brought a little something along for the ride – a road trip mixtape! Add these songs to your exploration playlist to get into the groove.

Elih, Bloom – 'The Long Run'

Yaima – 'Gajumaru'

Bob Marley & The Wailers – 'Jamming'

Nhii – 'A Mindful Blues'

Kevin Morby – 'Harlem River'

Sylvie Kreusch – 'Walk Walk'

Empire of the Sun – 'Walking on a Dream'

Iron & Wine – 'On Your Wings'

Caravan Palace – 'Moonshine'

Traveling Wilburys – 'Heading for the Light'

MØ – 'Way Down'

Parov Stelar, Graham Candy – 'The Sun'

Electric Light Orchestra – 'Last Train to London'

The Beatles – 'Mother Nature's Son'

EXPLORE YOUR WILD SIDE

You won't have to venture far to see that there are kind people, adventures to be had and views to be enjoyed everywhere. It's up to you to create your own experiences with people, places and circumstances.

Your wild and free spirit can show you exciting new ways that move you, amaze you and make you feel endlessly alive, and it does this best through the act of exploring. This word 'exploring' covers anything from travelling and trying out new things to opening up to opportunities and living in the moment. All these experiences are allowing you to make new memories and learn new lessons along your life's journey, which means the act of exploring holds the potential to change your outlook and can help you see things in a new and different light.

Exploration can be done anywhere – from the untamed wilderness of nature to the modern world erected by humankind. I will go into the best of both worlds, as I've discovered that both offer a chance to reconnect with what makes us feel wild and free. Even so, I will gradually shift our journey into the direction of nature, as this is the place we often don't add enough significance to.

Going forward, I will mainly talk about exploration in the more outdoorsy and adventurous sense. But feel free to use the lessons of exploration in all areas of your life to make the most of the opportunities and events that unfold.

IF NOT NOW, THEN WHEN?

IT'S ABOUT TIME

You are alive today, in this hour, in this minute, in this very second. How much time you have left is unknown, which means time is of the essence. Being aware of the fact that time is fleeting, you can learn to differentiate what is worth your time and what's a waste of it. Think about how precious time is for a minute. Really think about it. Try asking yourself the following:

☮ How can I make the most of my time?

☮ What can I do to stop wasting time?

Now these are some of the questions that I've kept asking myself on a regular basis for over a decade. The answers have driven me to explore as much as I could, because I thought to myself, 'If not now, then when?'

I believe there is significance to be found in each moment – it can range from the exciting to the simple, from the exceptional to the mundane. I open myself to the beauty of it all by seeking excitement on a regular basis, but also practise to revel in moments of simplicity.

Every second, time is handed to you, and you decide what to do with it. At this very moment you can decide what you do for the rest of the day. Will you catch the sun or reach for the moon? Will you try your hand at something new or set aside some time just to be? Or will you get a hold of your to-do list or lend a hand to someone else? When you have time on your hands, you have

79

endless possibilities at your fingertips. All you have do is to reach out, take action and seize what is yours.

Rise and shine

As the sun returns from the darkness of the night and is reborn on the horizon, so are you. A night of sleep is how you reset, and a new daily cycle begins the moment you open your eyes in the morning. Even your physical body is constantly renewing itself, as the body replaces most cells every seven to ten years. Some cells are revamped even faster, like red blood cells that have a lifespan of a couple of months. This means you'll never be the exact same as you are right in this moment again.

You can use your constant transformation and renewal to see each day as a new chance. Every time you wake up, you can make a habit of thinking to yourself, 'It's a new day, and I'm a new me. I'm not the same I was yesterday, and tomorrow I will not be the same as today.' This thought opens you up to continuously grow, change and evolve in ways that come natural to you because you don't cling to one idea of who you are. It's time to stop fighting the natural process of change and allow yourself to just be, without limiting yourself or your potential.

From this point onwards, you can leave behind past versions of yourself and step into your present wholeness. I encourage you to begin today, continue tomorrow and never stop.

From ritual to routine

Daily reminders can make us rise and shine, but they can also be the very thing that turns a rare ritual into an everyday routine. Ideally, exercise, meditation and contemplation shouldn't be rare occurrences, and neither should other things that benefit your mind, body and spirit. Now, to be honest, I've often found it difficult to have a routine, especially on longer nomadic trips. I would find myself waking up in a new place almost on a daily basis, and it took time and energy to adapt to my new surroundings. I took that as an excuse not to practise

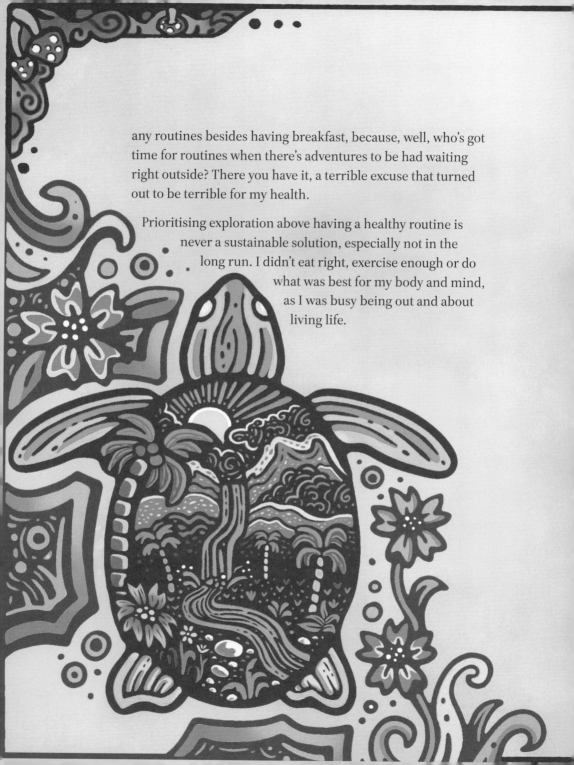

any routines besides having breakfast, because, well, who's got time for routines when there's adventures to be had waiting right outside? There you have it, a terrible excuse that turned out to be terrible for my health.

Prioritising exploration above having a healthy routine is never a sustainable solution, especially not in the long run. I didn't eat right, exercise enough or do what was best for my body and mind, as I was busy being out and about living life.

But as my health declined and my energy drained, I started learning my lesson. I began taking the time for myself to have a good routine, and it didn't take long to see that it worked wonders. It positively affected me both mentally and physically, and the effects became even more noticeable as the benefits carried out into the rest of the day. Now, I believe that wellbeing should be the first and foremost priority, always – there's never an excuse to neglect it!

Begin making healthy everyday decisions and change your habits to create a routine that benefits your wellbeing. It's commonly said that it takes on average 21 days to build, change or break a habit, but more recent studies suggest it can take between 18 and 254 days – on average it takes 66 days – to form a habit. This is just to give you an idea that if you want to create a routine that comes naturally to you, you have to make a habit of implementing it daily until it sticks.

You can start now by thinking about what your idea of an optimal daily routine is. After you brainstorm and evaluate, you can begin making a game plan. Remember that you want to implement your good habits on a daily basis and that you have to set aside sufficient time for this.

Everyday routine ideas

- Stretch: Wake up your body and notice the sensations.

- Body scan: Notice how every part of your body is feeling (see page 65).

- Grounding: Connect yourself to your present surroundings (see page 65).

- Meditation: Pause your thoughts, stay in the present moment (see page 42).

- Dream a little: Think about what you look forward to today.

- Write notes: Watch your thoughts and emotions, and write them down.

- Exercise: Stay active, move your body and get your pulse up.

- Pursue joy or purpose: Something that makes you feel uplifted or driven (see page 153).

- Bless your body: Affirm yourself with: 'I bless my body and my body blesses me'.

Slow down time

Hours, weeks, months and years sometimes seem to slip by, and sometimes time just doesn't go fast enough. Even though you aren't able to change the course of time, you do hold the power to change your perception of it. As you may have noticed, how a clock measures time and how you perceive it are quite different. Whether you're a fast-paced person or you like to live life in the slow lane, you can choose to slow down the trickling of the hourglass.

Stop overthinking

In moments where you get lost in thoughts of the past or worries of the future you rob yourself of experiencing the present. Try making a habit of paying attention to your surroundings, the smells and the atmosphere. What can you hear, taste and feel? What is the weather like? Are any people or animals around? Intentionally capturing these details will help you counteract the tendency to overthink.

Do something new

When you experience something new you automatically become more alert and present because you take in the situation. As your brain is writing down these new memories, you perceive the passing time to last longer. Try to do things that you haven't done before, explore places you haven't been and take opportunities that spike your curiosity.

Spend time wisely

Make a habit of putting down your phone. Don't consume too much of what keeps you hooked and reduce all forms of digital entertainment that rob you of time. Choosing how you spend your time is a way to knowingly live a more meaningful life. If you value your time you are more likely to pursue more positive experiences or find new ways to grow.

Romanticise your life

Each moment holds promise, yet when our ego is behind the wheel, we are very likely to go down a road of self-pity, resentment and negativity. Either we are envious of someone else or we romanticise far-off places, thinking that if we were somewhere else, everything would be fixed and our troubles would dissolve.

These kinds of thoughts are wasting our own time, so we need to have constant reminders to obtain and maintain a positive outlook that is centred around our own life and that has ourselves and our mental health as its main priority. Your everyday life is the place you can begin to knowingly shift your focus and start making a point out of experiencing every moment in the best possible way.

You deserve to see the worth of life in as many moments as possible, and you should notice its endless significance and countless wonders, because, believe it or not, they can be found everywhere you look – even in the mundane! In fact, it's the most important place to find significance, since your everyday life is where you spend most of your life. If you think along the lines of 'same old, same old', it can become very monotonous to repeat actions day in and day out and the chance to develop a feeling of nothingness is increased, but if you choose to explore the ordinary you come across the extraordinary.

Once you begin to open your eyes to the small wonders of your everyday life, you will learn that you have the power to turn moments from the ordinary into the extraordinary, the insignificant into the significant. You have the power to change the way you experience the world and all it entails. What once went unseen can now move you deeply within, and so you can build a deeper emotional connection to all that surrounds you.

Seize the day

- Have a good morning routine
- Surprise a loved one with a tasty picnic
- Gather your old t-shirts and do a tie-dye session
- Doze in a hammock
- Collect a bouquet of wildflowers
- Bring to mind five things that you are grateful for
- Go on a foraging quest in your local forest

Revel in the Moonlight

- Do some stargazing
- Create a theme night and match your food and music
- Go on a midnight hike and bring some herbal tea
- Listen to a comforting bedtime story or podcast
- Rewatch your favourite childhood movie
- Create a dreamy bubble bath by adding flower petals and lighting candles

89

Rest and rejuvenate

Do this, do that. Do everything all at once. Do it today, tomorrow or the day after – that's the schedule for many of us, maybe it's also like that for you. You might know what it feels like when there are endless things to do that make the mind numb, the body weary and the spirit low. You might not even prioritise rest and only get some every now and then.

This is a little wake-up call telling you that it's absolutely vital that you keep your drive strong and your life force flowing, and you can only achieve that when you regularly take time to rest and relax. You can only regain vitality when you do absolutely nothing, or, at the very least, engage in relaxing activities that let you chill out, release tension and ease your mind. One way could be dozing in a hammock, swaying side to side without a care in the world with your favourite tunes playing in the background. Just imagine that – what a vision! So simple, yet you might not set aside time for this, even if you truly want to.

You might be thinking, 'Just one more hectic week and then things will quiet down', and if that is the case, then ask yourself how many times have you told yourself this and if you've ever gotten to the quiet part. This is what I started asking myself when my art business started taking off, which was also while we were converting our van. One week after the other, I told myself that things were about to quiet down, until I realised that I kept being proven wrong.

Thinking back, it was an absolute mess, but at the time we thought we could make it work. What happened instead was we reached the point when everything started boiling over. The

plans that seemed so simple had become a jungle of stressful to-dos, and we were even sleeping in a construction site, always reminding us that we were still far from done. My body began showing symptoms of my own neglect, I started feeling pains and aches physically, and mentally I was breaking down on a regular basis. I realised I had reached the bottom of my energy pool and drained myself of all my life force. Some of the health struggles, which I still have to this day, are a result of me draining all the energy I had. That's when I learned I would have to deliberately set aside time in my busy schedule to rest and rejuvenate. Only then could I replenish my energy levels.

So take it from me, whatever passions you have, however much they enrich you now or promise you happiness in the future, you cannot allow yourself to be consumed. A hectic schedule and endless to-do lists build up stress, even if you are having fun most of the time. You need to tend to other aspects of yourself that give you joy, that can help you take your mind off things and make you relaxed. Don't make excuses for why you should postpone resting; instead, make excuses on the behalf of your wellbeing for why you should rest now.

Find your inner place of calm

Follow these steps while in a comfortable place, and navigate yourself to your inner place of calm. Your own personal sense of what brings you joy, calmness or relief will help you get there. Once you have discovered it, you can revisit whenever

you need an escape or want to regain control of the course of
your mind. This exercise can also be done as a halfway point to
bring stillness to your mind.

1 Close your eyes, clear your mind and focus
   on your breath. Think, 'I am safe. I am calm.
   I am present'.

2 Imagine following a pathway through different
   natural scenes. The path could be leading you
   through a lush forest, past a wooden cabin
   or down to a secluded beach. You can stop
   whenever you feel you have reached a place
   where you feel at ease.

3 When you have found that perfect spot, have
   a look around and bring it to life by imagining
   how you would sense it. What can you hear,
   smell and feel?

4 Allow yourself to stay in this place as long as you
   need. You can even imagine taking a stroll around
   the place or going for a swim — it's up to you what
   you do here, as long as it makes you feel good.

5 When you are ready to return to reality, imagine
   walking back step by step. Notice how your
   body feels right now. Notice your thoughts and
   emotions. Slowly open your eyes and stretch
   your body.

WONDERFULLY WILD

What makes you dream of the non-existent? What allows your
heart to speak? And what makes you laugh and jump with
joy? It's your imaginative, creative and playful nature. When
they are unleashed and able to flow, they show you the wild
side of yourself, as you become resourceful and spontaneous,
and start to take chances and live a little. Often, the three of
them overlap in some way or another, meaning you might
use your imagination to be creative, playfulness to imagine
and creativity to play. Believe it or not, they have the potential
to create real-life miracles. Through your imagination, you
can enter a new world of ideas. With creativity you can create
something out of nothing, and by immersing yourself in play
you can be uplifted by positivity.

Your imaginative nature

Imaginative thoughts are the starting point of visionary ideas, enticing stories and fantastic masterpieces. Using imagination can bring forth creation, innovation and change. If it wasn't for humans' ability to imagine, the world would not be the same. Close your eyes for a moment and think about it. How would the world look if it wasn't for our ability to think of new things? And just like that, you used your imagination.

Imagination allows you to envision everything, even that which doesn't exist, meaning imagination is not just a blessing, it can also be a curse. For example, it's your imagination that conjures up worrisome images and fuels fearful thinking when you imagine worst-case scenarios.

This is one of the moments your imagination can head in a self-destructive direction, but you can practise to maintain control of your imagination so it doesn't just wander off.

You can change the course of your imagination to anywhere that you please, so why not lead your thoughts in the direction of calmness, happiness and serenity? Or maybe use it to manifest success and materialise goals (see page 152). Imagination can be a means for you to choose the course of your thoughts and the state of your mind, and both play vital roles when deciding the following actions that you want to take.

Activity: Practise imagination

Imagination is, in its purest state, a complete escape from reality. Here anything is possible – nothing is restricted by reason or has to follow physical laws. When imagining something it helps to close your eyes, so your visual sense is not tied to the material world. It takes some practice triggering imagination, but it helps to first imagine something that does exist and gradually move in a more abstract direction, which can defy all logic and reason. For example, if you want to imagine yourself in another reality, it helps first imagining yourself as you sit in whatever place you find yourself in at the current moment. This makes it easier to imagine how you would move out of your reality into a different one. You can imagine how you would stand up, put on shoes and step outside. Then you could walk down an alley or path that has a portal at the end. It might take a few tries to get there, but once you can imagine this process, your imagination can begin to take over. Imagination can be a playground for you to explore. Here you can create new ideas and visions that you can take back and apply to the physical world.

Your creative nature

Creativity is an immense force that has the ability to make significant changes happen on both a personal and societal level. On a personal level, the list of wonders includes the ability to improve mental health, as it can be a therapeutic way to de-stress, overcome past traumas and be a form of self-expression. On a grander scale, creativity is the source of development, change and progress. With creative actions, the visions of your imagination can turn into reality, making the impossible possible.

You don't need to be an artist to call yourself creative, and you don't need to be talented – creativity isn't about masterpieces and perfection. You don't even need to work towards a specific end result – you can be creative just for the sake of it. Creativity is simply creating something – there are no rules, no methods to stick to, no rights and wrongs.

When you find ways to be creative in your daily life, you get to harness its infinite power. Even if you don't count yourself as creative yet, this can change. There are so many means to unleash creativity, like painting, crafting, writing, sculpting, building, sewing, woodworking, cooking, dancing, acting, playing and music just to name a few.

Activity: Create a natural mandala

This challenge takes place outside to put your creative nature to the test. Don't worry, you won't be graded! Do this in a place where it isn't too windy.

1 Find an object that will serve as the centrepiece. It can be a rock or a flower. Place it on an even spot, where there is room around it for your mandala.

2 Gather four twigs that are somewhat straight and the same size. Place them in equal distances around your centrepiece. For more intricate designs you can use more twigs if you like.

3 Now gather leaves, flowers, rocks, acorns and other natural elements around you. Almost anything you find in nature can be used in your mandala.

4 Begin placing the objects in circles beginning from the centre and moving outwards. There are no rules here – you are not striving for perfection but for resourcefulness and creativity.

5 Once you are finished, take some time to take in the work of art that you created. You can either let this art be or destroy it, as a way of celebrating the impermanence of life.

Your playful nature

Life is not to be taken too seriously. I mean, why should you? Eventually we all end up in the same place, and we might as well let loose and have a little fun while we are alive. Playing games, being silly, cracking up on jokes, moments like these can make your day, and even your week. Not only is your mood improved, but your overall health is too. Laughter releases tension and reduces stress because it triggers the release of endorphins, your body's natural feel-good chemicals. That's why the saying 'laughter is medicine' is no joke. There's never a good reason why you shouldn't have a little fun, even if you are going through a tough time. Maybe allowing your wild side to come out to play is just the thing you need to dissolve your worries.

Playful ideas

☮ Challenge a friend to a game of real-life bingo: items on the bingo sheet can include things like pet a cat, bake a cake, do some stargazing, find a cool rock, make a bonfire, go skinny dipping and pick wildflowers. The possibilities are endless, so let your imagination run wild. Think of a prize for when all items have been found or completed.

☮ Make a to-play list and add these items: build a fort, play hide and seek, play boardgames, do a puzzle, play minigolf, build a sandcastle. Tick off one thing at least once a week and add new as you go.

Step outside your comfort zone

Living on the wild side is easier said than done. There is always something within you that might make you want to stay put or do the things you know because it feels safe. Comfort is where the spoilsports live; you know, the guys that bestow doubts as they feed your ideas of logic, reason and responsibility. They can be such a drag! If you stay in your comfort zone and listen to the spoilsports, you never will see the amazing things that exist outside of it.

Ask yourself the following questions: How can I really get to live a full life, if I only do what I already know? How can I learn from my experiences if I'm always prepared for what is coming my way? What is really the worst thing that can happen if I let go of fearful thinking and leap into the unknown? Maybe don't answer the last one, because those worries are usually what the spoilsports get off on to create all kinds of excuses. Actually, try to forget the idea of the worst-case scenario altogether – your mind has an all too vivid imagination when you give into fear.

Instead, try to focus on the best-case scenario (see page 37). The more you use that as motivation to actually go through with stuff, the more you will see that what worried you before wasn't at all the actual outcome. The more you leave your comfort zone the more natural it'll become, and you learn to live more in the present instead of worrying about scenarios that will never happen. Once you embrace discomfort, you

see what you're really made of, and you might realise that you haven't been living up to your full potential. Thoughts or ideas you held about yourself or life will be disproven, limitations exceeded and barriers broken down. To quote Neale Donald Walsch, 'Life begins at the end of your comfort zone'.

My mixed bag of discomforts

Speaking from personal experience, the times I left my comfort zone I went through life-changing experiences, gained new perspectives or got new, cherished memories. My mixed bag of discomforts holds both light and dark stuff. I've travelled to places that intimidated me, even when some people advised me not to go. I visited spiritual communities and meditative retreats with no idea what I was getting myself into. I've challenged my fear of heights by climbing and abseiling, and once I even walked over a broken hanging bridge over a raging waterfall (I don't recommend that last one). I've made an effort to understand the history of the places I've travelled, learned about colonialism, genocides and war. I've faced the atrocities of animal agriculture, which eventually led me to change my whole diet. I've relied on absolute strangers while couch surfing and hitchhiking. I've started my own business and turned my passion into a career. And I've said yes to opportunities that intimidated me, like writing this book. For me, discomfort is a force for growth and change and it keeps helping me become stronger in countless ways.

The best of two worlds

Even if I believe discomforts are worth having in all kinds of variations, I will mainly focus on adventurous exploration going forward. Thinking back on all my experiences, I've discovered two big sources of discomfort that have made for the most wild and free experiences.

The first source is road trips, and here it doesn't matter if you are driving on your adventure or if you are hopping on a bus or hitchhiking. Once you are on the road, bumps occur, wild coincidences happen and amazing views are to be had. Roads always lead to adventure and ways to feel discomfort. The second source is nature, because nothing here is laid out for luxury or convenience. Here you are exposed to the elements, faced with the intensity of nature's force and all you can do is deal with whatever you are faced with and make the best of whatever happens.

If you combine the two – road trips that take you into nature – you have a potent source that guarantees you to feel wild and free. Not only do they have you overcome fears and obstacles, but you will experience the beauty of simplicity, as you can't bring a bunch of stuff with you. When you aren't surrounded by an excess of items, you are less distracted, less comfortable and less entertained, which means you can be wilder and freer to live in the moment.

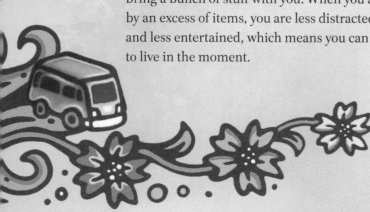

Affirmations to be wild and free

No matter what comes my way, I stay open,
free and optimistic.
I explore opportunities, even when they take
me outside my comfort zone.
I free myself from my own limitations and
fearful thinking.

EXPECT NOTHING – APPRECIATE EVERYTHING

We are living in a time of social media, which means we are constantly bombarded with snippets and snapshots of how someone else is seemingly living their best life. Even though the posts and videos could serve as inspiration for our own adventures, it's just as likely that they'll make us feel depressed. We might not have the time or money to do all the things that we want to do, which may make us feel like we are missing out.

So, how can you counteract this fear of missing out? Well, the answer is in the title: expect nothing – appreciate everything. If you go through life with this motto, you'll live your own reality and get to see the beauty of it. When you don't compare your experiences with others', you won't create impossible expectations for yourself. This leaves you with more headspace for ideas and plans that lead to your own exploration.

Expect nothing – appreciate everything is one of my favourite mottos in life. It teaches the lesson that sometimes you have to compromise, and that's okay. You'll be met with challenges that'll change your plans, and that's okay, too. You may even have to end your adventure and head home, and yes, you guessed it, even that's okay. The less you expect things to go according to plan or to match your expectations, the more room you create for serendipity.

Turn your bucket list into a nugget list

I am not convinced that the conventional idea of what goes on a bucket list is doing much good. I used to fixate on specific travel experiences, thinking I must see this or do that to feel like I am making the most of life. I felt disappointed whenever I didn't or couldn't check an item off, due to time, money or bad weather, and even if I did tick something off my list, my high expectations and meant the experience often didn't live up to them.

Like the time we visited the Plitvice waterfalls in Croatia, which we had seen many photos of online. I couldn't wait to take in their beauty, but when we finally got there, I was underwhelmed. The reason simply was that they looked spectacular from an aerial view, which is an impossible perspective to get unless you fly over or have a drone. Having the viewpoint of a hiker, the waterfalls just didn't look as significant and on top of that the tourist crowds made the place feel less pristine. But the real problem was my mind – if I hadn't placed so much pressure on the experience, I could have made the best of it.

That's why I started thinking in different ways and created an expect-nothing-appreciate-everything list as an alternative to the bucket list. However, this turned out to be a mouthful, so I changed it to the nugget list (I hope it catches on). This list doesn't focus on specific sights, places and activities, but rather broader aspirations, hopes and dreams that can be found randomly in all shapes and sizes. This has helped me appreciate whichever golden nuggets I find on my way and has broadened my capacity to explore and experience in a way that opens myself to the unknown.

The nugget list doesn't have to be limited to travel; items can be anything that have the ability to give you an authentic experience that invokes a deeper sense of fulfilment. I like to keep the list short, and update it when I feel I've obtained a golden nugget.

My nugget list

☮ Explore my current surroundings by taking time to wander around aimlessly.

☮ Have intimate experiences by opening myself up to deeper connection and receptivity.

☮ Discover hidden gems by listening to recommendations from locals and see where I land.

Activity: Make a nugget list

Consider your deeply held aspirations for what emotions you hope to experience on your exploration – remember to think in broad terms – and then write them down. This is how you can avoid fixating on specific outcomes by manifesting your general ideas of aspirations. Whenever you are about to head out and explore, remember it's not about filling your bucket, it's about the appreciation of coming across a nugget.

How to bring your A-game

Life is unpredictable – that's the beauty of it! If we start lowering our expectations, we can take life as it comes and make the best of it. The destination isn't what matters, it's the journey that counts. When you see life like a game, you can begin to play.

Allow your mind to open.

Accept what comes your way.

Adapt to how the events unfold.

Acknowledge the lessons that come to you.

Appreciate the game that is life.

Leave it up to chance

When you leave things up to chance, you can truly feel free and just enjoy the ride. I invite you to fully embrace the unpredictable and see how the events unfold. The next time you have time on your hands and you want to do something exciting, bring in the element of surprise. Instead of going into planning mode, try to do the opposite. The less you fill your time with a plan, the more time and space you create for serendipity to unfold. When you leave your plans up to chance, you open yourself to experience the rush of the unknown.

Give it a shot: When it comes to a single activity, only check out basic information about it. It can be a hiking trail, cultural sight, an event or even a movie. Go there without reading reviews or looking at photos. In this way you can experience it without any preconception or expectations.

Take a gamble: Wake up before dawn and take a hike to watch the sunrise. Not every morning creates amazing sunsets, but taking this risk is part of the gamble. Remember to expect nothing and appreciate everything (see page 106).

Rule of thumb: The next time you want to go somewhere, try hitchhiking. This activity unfortunately has a bad rep, but don't let that discourage you. The rule of thumb when it comes to hitchhiking is to follow your intuition and go with your gut feeling. However, I recommend you do this with another person just to be safe, and do check out the local laws beforehand, as hitchiking is illegal in some places.

Go all in: If you have a couple of days or weeks and you want to fill them with excitement, try leaving the whole itinerary up to chance. You can just pack a bag with necessities and venture out. It doesn't matter if you are on foot, do a road trip or board a plane. The point is living minute by minute, hour by hour, day by day.

The element of surprise

I can personally vouch for the fact that inviting in the element of the surprise can create instant portals to adventure and excitement. Some of my most cherished memories happened by chance. My favourite anecdote was the time we picked up Wuangsi, a Malaysian guy that was hitchhiking for the first time ever, who ended up travelling with me and Lenny for 3 weeks in our tiny van. At some point, the three of us had a business idea and headed to Berlin, the start-up capital of Europe, to see if we could make it happen. It failed, but, hey, at least we tried and had a bunch of memorable times with our new friend.

Creating space for chance can work small and big wonders you never even thought could be possible. It really proves the fact that some of the best things in life aren't those that were carefully planned.

RECONNECT WITH NATURE

Wilderness is your birthright – you are born to be wild. The reason for this is simple: everything in nature is wild, and being a part of nature, you are no exception. This is worth remembering, as it can sometimes be a little difficult to feel a connection to the natural world, especially in this day and age. Living a modern life, you might rarely affiliate with nature – you live in a house, buy food from the grocery store and have endless water coming from a tap. These are just some of the ways you have become estranged from the natural world, and if you think about it a little, you will probably come up with dozens more.

Now, I'm not here to tell you to give up your modern life to move into the woods, but it's worth reflecting on how you are disconnected, because this can give you some ideas on how to reconnect and maintain your ties with nature on a regular basis. One way can be to connect to your local environment and its flora and fauna. Try to learn about the nature where you live and find out which plants, trees and animals live here. You can start right at your doorstep, even though you live in an urban environment. Nature exists (almost) anywhere – we just have to look for it.

Naturally changed by nature

Just by stepping into nature, you are exposed to nature's green energy, and changes can begin to happen naturally within. It has been proven by research that spending time in nature positively influences your wellbeing. Studies show that there are at least 227 non-material ways that nature is contributing to and benefits our wellbeing.

Here are some of the ways nature restores and improves both our mental and physical health:

- ☮ Increases feelings of calmness.

- ☮ Increases feelings of joy and happiness.

- ☮ Increases a sense of belonging.

- ☮ Restores energy levels.

- ☮ Improves capacity for concentration and attention.

- ☮ Reduces symptoms of anxiety and depression.

- ☮ Reduces feelings of isolation or loneliness.

- ☮ Reduces irritability and tension.

- ☮ Lowers blood pressure and stress levels.

- ☮ Boosts the immune system.

Strengthen your bond with nature

Reconnecting with nature will likely take longer than a couple of hours. This should come as no surprise, because when you inspect modern life with all its comforts, nature seems almost the opposite of the world we have created to live in. So, when you head back to nature, you need time to reconnect. The way I like to think about it is that nature is like a really good long-lost friend. You need time to catch up to really get to know them again. Reunited with nature, you need to acclimate to its environment before you can fully immerse yourself in it.

Now, what are you waiting for? Nature is out there waiting for you. To get there you have to turn an idea into a goal and then make a plan. You need to set aside time to decide when you'll go, where you'll go and how you'll go. I will try to help you with this planning process and motivate you to try out some new things.

Right now, you might not count yourself as an outdoor enthusiast, and you may not feel an urge to leave your comforts to pursue the wild. You might have other priorities, thinking your time will be better spent somewhere else. Perhaps you're so used to city life that going into nature seems like an unnatural undertaking or that it's something reserved for 'sometime in the future when it's more convenient'. If any of this comes close to how you are feeling, don't fret. I've often felt disconnected to nature or prioritised other activities, but when I was reminded of what I was missing, I suddenly felt a pull back. As I found my way back into the wild, I noticed how much I'd actually needed nature time. This has led to making more plans that brought me back into the wild and kept my bond with nature strong.

On the move

I like to strengthen my bond with the wild and get closer to nature, by literally moving out and about. From walking and cycling to rolling and paddling. By simply moving, you can experience nature up close, and in this day and age, there are tons of ways you can move from one place to another. Some offer up-close or scenic encounters with nature, others allow you to reconnect more closely with your element, aka your body (see page 55).

My favourite experience of being on the move was during a three-day kayaking trip in Laos with overnight homestays. We could only bring a couple of essentials, and otherwise spent the days paddling past limestone cliffs, elephants and quaint villages. The trip was 120 km long and ended in the capital Luang Prabang. Being so energised by this experience, we followed it up by renting bicycles at dawn and riding to the Kuang Si waterfall, and beat the tourist crowds that later arrived by tuktuks. It was really nice being at such a pristine spot and enjoying nature without having to share it with other people.

GO CAMPING

Would you like to see nature transform during the night and take your time to watch the stars? Do you want to be able to enjoy the simpleness of a night out in the wild, and feel it slow down your life? Camping is a way to get back to your basics. Experiencing this kind of simplicity not only allows you to appreciate the little things in life, it puts everything else in your life in perspective. You might realise what really matters to you, and it can bring to surface how you feel deeply within.

When you're out camping and facing difficulties, such as a change in the weather or a leak in the tent, you can use this challenge to remind yourself of the reason you're here. This opens you to accept all aspects of camping and allows you to grow from its simplicity, even if it's not always easy or comfortable.

121

Escape the hustle and bustle

Spending time at a spot with a simple outlook can slow down your life. First things first, you have to be resourceful and look at what options are within your immediate reach. Most of my trips were done on a dime, so I am talking from personal experiences when I say you don't need a lot of money to experience nature and make it worthwhile. You can simply throw a mattress in the back of a car and hit the road, or ask your friend to borrow their old tent. Look for opportunities and take them. The more basic the setup, the more resourceful you'll become while exposed to the elements of nature.

Lower your expectations for the perfect weather, perfect setup, perfect circumstances and perfect trip. If you go into this expecting perfection, you missed the point of focusing on simplicity and having appreciation for the little things. Getting out into nature is not about fulfilling your ego's expectations; it's about connecting to the expansive viewpoint of your wild and free spirit. This is how you take the natural course and mirror the wilderness that you surround yourself with.

Remember that the trip you are planning is about nature, so prepare yourself to put your phone down. You want to refrain from digital distraction while you are spending this precious time outside – don't browse the internet, don't use social media and don't email or text. This is your chance to communicate and reconnect with nature or to go more deeply within your own wilderness.

How to stay in the wild

Camping is how you can get up close, immersed and connected to nature. Even if you might not count yourself a happy camper yet, that can change. A little planning and mental preparation can go a long way and if you come prepared, you up your chances for a great trip.

Camping simply means sleeping and dwelling in nature, and is a way to get back to basics and to experience simplicity. Camping can be anything from hammocks and tents to luxurious RVs and everything in between. The many options allow you to find the type of shelter that suits you best. You can rent, borrow or buy, whatever works for you.

Here's a little tip for you: if you're considering buying one or the other, or even converting your own camper van, carefully consider if this is the right choice for you. You don't want to spend a lot of time and money on something that turns out to be a bust. It really pays to rent a tent or a rolling dwelling to try it out before investing in your own.

Exposed to the elements

You are at the mercy of the wildness of nature the moment you expose yourself to the elements. Chances are you'll feel more vulnerable, and it's very likely that you will be driven out of your comfort zone. Yes, nature can be intense to endure – it has a way of overwhelming you with its green force. The grounds and air are alive with life, which makes an endless spectacle for the senses.

We live in a time where we can ease into the intensity with the help of good, reliable gear. Whatever your needs are, there is a gadget to cover that need. You can generate electricity while being out in nature with solar power, purify stream water with ultraviolet light and water filters, protect yourself from insects with all kinds of nets and sprays and even bring a portable foot massager to help you wind down. Everyone has a different threshold to how much they can endure, so I say bring whatever you need to make that trip happen.

Mindfulness in the wild

Once you arrive in nature and immerse yourself in this lush environment, you might think to yourself, 'What now? Is this it? Is this what all the buzz is about?' Yup! It's as simple as that. In the sense of camping, I don't believe nature is here to serve you with distraction, entertainment or excessiveness; it's here to open you up to the rawness of life, making you feel bare and

exposed to forces bigger than yourself. This will bind you to the moment of now, instead of being lost in thoughts about the past, or projections of the future.

Camping is therefore a means to mindfulness, but in more ways than the common understanding of the word. It starts by packing, as you become mindful of what is really important for you. By only being able to bring a handful of things and a couple of necessities, you have to prioritise what really matters. Going camping is a way that you can get back to basics. Living in a state of simplicity you might see that all the excess you surround yourself with at home is probably unnecessary, or the quiet might reveal something you had not paid attention to in your life.

Getting out into the wild is a way that you can raise awareness in a number of ways, giving you a truer sense of yourself and your reality. Having a more expanded perspective is what can make you more mindful in other aspects of your life, making you contemplate what you think, do, eat, consume, possess and accumulate. It can also help you see the ways you've become disconnected, be it to nature, people or even yourself.

Leave no trace

Wherever you're going, you need to be aware of the environment you're heading into. This is how you can plan ahead and prepare in ways that respect the wildlife and minimise your impact.

It's important that you set aside your own convenience and go out of your way to avoid doing anything that could result in an imbalance, destruction or depletion.

In the forest, at the beach, in the mountains, at the lake, in the desert or over the hill – no matter where you find yourself in nature, you need to live by its laws: respect its premises, follow its principles and maintain its natural order. Your planning and consideration go a long way to protect natural habitats and its inhabitants. Even when you are at home and are taking everyday actions, your decisions impact nature (see page 187). The more you keep it natural, the more you can reduce your environmental footprint.

Leave no trace principles

There is a common set of principles that responsible campers live by to do their part for nature. It's called 'Leave no trace', and revolves around the idea that the only thing you should leave behind are your footsteps. Of course, there is more to it:

1 Plan ahead and prepare – learn about the place you are going and pack accordingly.

2 Travel and camp on durable surfaces – don't alter terrain and opt to stay in designated spots.

3 Dispose of waste properly – bring reusable bags to take trash with you, dig a hole to bury your droppings.

4 Leave what you find — what grows and belongs in nature, shall stay here. Don't make alterations.

5 Minimise campfire impact — use established fireplaces, keep fire small and scatter the ashes when cool.

6 Respect wildlife — don't disturb animals, watch them from a distance. Don't approach or feed them.

7 Be considerate towards other visitors — be courteous of others. Let nature's sounds prevail and don't be loud.

Activities in nature

There are plenty of activities to be enjoyed throughout the season – make sure to make the most of your natural environment throughout the yearly cycle.

Spring activities: Pick flowers, create wreaths, go birdwatching, plant seeds, fly a kite, ride a bike, have a picnic, watch a sunset, walk on the beach at low tide.

Summer activities: Dance in the rain, walk barefoot, relax on the beach, look for wild berries, take a nap in a hammock, do a yoga flow, rent stand up paddleboards.

Autumn activities: Go on a foraging quest, brew tea to enjoy outside, practise skipping stones, jump into a pile of leaves, watch a sunrise, go mushroom spotting.

Winter activities: Play in the snow, make a snow angel, go sledding, go ice-skating, take a snowshoe hike, build a snowman, do some stargazing.

130

BE AT ONE WITH NATURE

Aligning myself to nature has become an elemental part of my everyday life. I usually find this easier when I am in a forest or in the mountains, as this is where I find myself really vibing with nature. I don't know whether it's the crisp air, the green surroundings or the fact that all that can be seen, heard and felt is pure nature. These are the places where I feel I truly belong and am at one with the elements. Nature is the place where I always feel that every minute is well spent. My senses become heightened, which brings my mind to myself and my immediate environment. That's how I can make the most of each moment and each season, from spring and summer to autumn and winter.

The more I am out in nature, the more I feel at one with it. This is where I'm faced with the fact that I am made of the same elements and composed of the same matter. I drink the water, breathe the air, and find nourishment through the soil. I keep rediscovering that I am in fact nature, no matter how modern or urban my life gets.

Come to your senses

Our busy modern life often has us disconnected from ourselves, which is made noticeable by the fact that we are not paying attention to most of our senses. Your senses are the very means to experience reality, help you gather information about yourself and your environment, but there is one sense we do use excessively, which is our sense of sight, that can keep us so captured that our eyes become glued to screens. And so we binge, scroll and repeat as we overstimulate ourselves with digital content.

Some say we have 21 and even up to 53 senses, but I will keep it to ten senses, because what is more important is that you learn to become more mindful of what you can sense. At any given time you experience several senses at once that combine to create your perception.

Apart from sight, hearing, smell, taste and touch, your senses also include your sense for balance (equilibrioception) and temperature (thermoception). You also have a proprioception (the sense of position and movement of your body), which creates body awareness through receptors in your muscles and joints. Then there's nociception, which is the perception of pain, and our sense of interoception which helps us take note of the signals and sensations we feel within the body.

Sensory walk through the forest

You can try to pay closer attention to all your senses to help raise your inner awareness. Try this activity to activate them, which can also be seen as a kind of mindfulness meditation through a forest. I encourage you to take off your shoes to get connected with the earth under your feet. As you walk, calm your breathing and let go of thoughts that distract you from this moment. Try to acknowledge each of your senses. Take breaks in between to become aware of each one.

Vision: Use your sense of sight to take note of your surroundings. Begin listing what you can see, like plants, trees and objects. Notice how as you move, your perspective and viewpoint changes. Note the depth of field.
Can you see the horizon?

Audition: Use your sense of hearing to pay attention to the soundscape. What can you hear? Try to make sense of each individual sound. Cup your ears in different ways to notice how the soundscape changes.

Tactile: To engage your sense of touch, use your hands and feet to feel the trees, their bark, the leaves and the ground. How does each feel?

Thermoception: Tune into your sense of temperature by noticing how different matters and elements feel. Which feel cold and which feel warm?

Olfaction: Use your sense of smell to take in the scents of the trees, plants and earth. Take a leaf and crunch it to release the scent from within. Do so with flowers and even try smelling a rock.

Gustation: To engage your sense of taste, stick out your tongue and try to taste the air. Can you taste the forest air? You can also scour for acorns or berries – just beware not to eat anything poisonous. Close your eyes to become more aware of the taste.

Proprioception: To connect to your sense of movement (also called kinesthesia), close your eyes and walk a few steps. Notice when there is a change of terrain your body is using your muscles differently.

Equilibrioception: Walk along a fallen tree trunk and notice how your sense of balance becomes more noticeable as you carefully walk on it.

Nociception: The sensation of pain is a sense we like to avoid, but you can carefully activate this sense by walking barefoot on terrain that is uncomfortable, such as gravel or rocks.

Interoception: Connect to your inner sensations by closing your eyes and asking yourself, 'How do I feel?' Are your basic needs met, or are you hungry, thirsty or tired? Is your heartbeat regular or fast?

Journey beyond

It's time to go the extra mile and go even further than you have ever ventured. I invite you to think big. In fact, I invite you to think bigger than yourself, bigger than this planet, even, and think as big as the universe. Because when you think big, the impossible can become possible – you can reach your goals, live with purpose and find more meaning in life.

Thinking big allows you to focus your attention outwards on other people and the world, which is necessary to be connected to everything and live in loving unity. Yes, when you journey beyond, the sky's the limit, and you can make big things happen – for yourself and the world.

This part is about the extra effort that you can make that not only allows your wild and free nature to continuously thrive but also allows you to reach your full potential. By opening yourself to a continuous journey of growth and evolution, you allow your free spirit to take the lead.

The journey beyond is one I've taken many times. If it wasn't for my willingness to keep transforming myself into the person I aspired to be, I couldn't have made my travel plans come true or my dream profession a reality. I wouldn't have been able to change my bad habits or let go of people that were not good for me, and I wouldn't have taken chances that were not always comfortable or easy. Instead, I would have kept bad patterns and habits, and my aspirations would have remained wishful thinking. Basically, I would simply not be the person that I am today.

INVITE NATURE IN

Your environment is the physical space that surrounds you. As you've been out and about exploring nature, you might have felt its lush environment transforming your mood and wellbeing. Feeling the good effects, you may also have developed a natural craving to spend more time outside, but, as most of us live sheltered lives in cities and work 9–5 jobs, it might not be possible for you to head out into nature on a daily basis.

The good news is that there are ways to bring you closer to nature within your own home. One of these is biophilic design, where natural elements like plants and natural lighting and materials are brought inside. Not only are these aesthetically pleasing, but inviting in nature can improve your overall wellbeing. Opening your home to nature's powerful energies is, of course, no substitute to spending actual time outside, but it can help you feel closer and more connected to nature in your everyday life.

I've learned a calming, clean, decluttered and purified home environment makes a thriving ground for me to live. Not only do I feel more comfortable, but I feel the space opens up to unleash wilderness, which makes me feel more in touch with my emotions, creativity, imagination and playfulness. No matter if I am in my brick home or my home on wheels, I've learned that my wellbeing relates to my immediate environment and has a similar effect to that of my loving relationships; I can either feel safe, happy, inspired, uplifted and loved, or, when the environment is messy and uninspiring, overwhelmed, confused, agitated and disconnected.

Cultivate your environment

Coming up are different ways that you can welcome in the elements of nature to calm, soothe and heal you with its green force on an everyday basis.

Keep it simple

Having experienced the simplicity of camping, you might realise that all the stuff that is filling your home is also cluttering your mind. The more you own, the more you have to take care of, organise, keep clean and navigate around. Start removing things from your environment that you never or rarely use. The goal is to only possess things with a purpose, may that be functional or visually stimulating, otherwise your material attachments might disconnect you from what really matters to you.

Materials and atmosphere

Centre your living environment around the elements of nature and decorate with nature in mind. Bring in a good amount of plants to bring in greenery and add art prints, photography and objects that make you feel nature's presence. Avoid using materials that are synthetic, and instead focus on natural materials like hardwood, bamboo, rattan, cotton, cork, glass, stone and clay. Everything you incorporate should optimally be sourced from renewable materials. Always opt to buy things that are produced in an ethical way and have a low impact on the climate.

Colours

The colours you surround yourself with can influence your mood in different ways. Colours like red, orange and yellow are associated with the warmth or intensity of fire, which we perceive as comforting, creative or energetic. Colours like green, blue and purple are associated with plants, water and the sky, which can be motivating, soothing and help raise awareness.

We can also have emotional associations of colours that are rooted in culture, which can influence our perception. For example, the colour blue is often associated with melancholy, green with envy and indigo blue as royal. You can even have your own subjective experience of colours, which might make you prefer certain colours over others.

When you pick colours, you can guide yourself with the overall colour perceptions that I've listed below, but I advise you to follow your intuition and sense for yourself to see how the different colours affect you emotionally.

Blue: soothing, calm, tranquil, melancholy

Green: safety, motivation, healing, nature

Yellow: energetic, cheerful, optimistic, reflective

Orange: spiritual, creative, confidence, stimulating

Red: passion, power, intensity, provocative

Pink: kindness, nurture, romance, intimacy

Purple: wisdom, awareness, luxury, spirituality

Brown: stability, harmony, reliability, earthly

Black: mystery, strength, elegance, dignity

White: peace, quiet, hope, purity

Scents

Dried herbs, flowers, fruit peels, spices and essential oils
can bring the perfume of nature into our homes, and
you can create your own potpourri to use as a natural air
freshener. Look to see what plants are available to you and be
resourceful – you might have lavender, rose petals, rosemary
or peppermint growing in your garden or neighbourhood. To
dry them, bundle them up and hang them upside down, and
when they are ready, place them in a bowl. You can also add
dried fruit peels or whole spices, like cloves, cinnamon sticks,
cardamom and star anise, and choose a matching essential oil
to drizzle over. There are many combinations. My favourite
potpourri bowl is forest scented and includes wildflowers,
citrus peel, pine cones, nutmeg and pine tree essential oil. I
like to change my natural air fresheners seasonally.

LIVE INTENTIONALLY

Up until now I have covered a wide range of topics – everything
from mindfulness to exploration. What all of these chapters,
activities and lessons have in common is that they teach the
power of your free will. You are free to be who you want to be,
and starting with your intentions, you can make tiny changes
gradually that might eventually take on epic proportions.

The butterfly effect

In chaos theory, the butterfly effect is a term used to describe a concept where even the smallest of actions can create significant and unforeseen effects in the future; how something as small as the flutter of a butterfly's wing – through a chain of cause and effect – can create a storm on the other side of the world. This means every little action and every subtle change that happens in the world can influence and change the course of events.

On a personal level, it means what begins with a thought or impulse is carried out into your actions. One little thought affects your present and therefore onset actions that change your future. Because of this, everything you think, say and do has an impact, as one thing leads to another and to another, but you are never in full control of the course of events, as everything else in the world also has an impact on you. So the best thing you can do is to continuously align your actions to your intentions, because in this way you still set into motion what you wish to bring to the world.

Intention setting

If we intend to be happy, we might need to quit our job and pursue our dream career. If we intend to heal from the past, it could be that we need to apply means like instrospection or therapy to recover. When we have plans, hopes and dreams,

we have to make it our intention to pursue them. Intentions are what can help us guide our actions so that we can reach our goals and actualise our aspirations.

Intention is the starting point of actual change, transformation and elevation, because whatever we bring to mind can from here on out multiply and manifest within us and out into the material world.
We can have many intentions that span from ourselves and our relationships to other people and even the entire world. The more intensely we bring them to mind, the more focused we can set them into motion. In some ways, having an intention is like holding a seed.
In order to turn this intention into actuality, you need to incubate and nurture it. The moment you begin taking real action, your intention can grow and take place in reality.

I remember when I made my first intention of what I would pursue when I was grown up. At the age of 14, I decided that I wanted to explore the world. My intention was so potent that it stuck with me for a long time, and still lingers around today. I feel I've had my fair share of travels, and I've set new intentions that allow me to explore other aspects of my wild and free nature. Now, I get to be more creative and enjoy the simplicity of staying put. Even so, I think it won't take long until I set off for a big trip again, because that's just how my free spirit seems to roll.

Personal intentions

I encourage you to contemplate your intentions. Ask yourself, 'What do I wish for myself, for others and for the world?' Close your eyes and think of what comes to mind. After you have thought about this for a minute, write down what came to mind. You can start by writing: My intention is to ... Try not to think too much as you write, just pour whatever comes to mind from pen to paper. Punctuation and spelling don't matter here, what matters is that you feel free to write what comes naturally to you. This process is a way to gain more clarity about your hopes and dreams.

You can begin growing your intentions by doing a ritual. I've added a ritual of far-reaching intentions, covering the grounds of personal growth to intentions for your fellow earthlings and the world. Each includes a flower or herb that holds properties, making the intention even more potent. Remember, having an intention and creating a ritual is just the setting of the intention; after that, you have to take the steps to cultivate it to bring it into actuality.

Step 1. Choose your herbs or flowers

Rosemary – keeping both feet on the ground

My intention is to be in the here and now. I maintain a solid connection to earth. Thoughts, feelings and emotions move through me but don't have the power to waver me. I will not be thrown off balance. I hold my ground and stay present in this very moment. I am grateful and I practise deep appreciation for my life.

Jasmine – kindness and love

My intention is to always be mindful and kind. I give my help to those who need it. I share my bliss with others and enrich the beings around me. I listen, respect and consider other earthlings. I aim to keep widening my circle of compassion. I aim to stay impeccable with my word and true to myself. I am kind to myself and stay in touch with my emotions.

Rose – growth and transformation

My intention is to always rise to new levels of my being. I aim to progress by keeping myself open to new perspectives, new knowledge and new opportunities and know these are elemental to the natural process that is change. I learn my lessons and apply them to my life, knowing I'm always growing and constantly transforming into new versions of myself.

Lavender – healing and transcendence

My intention is to cultivate myself and bring forth healing through introspection and inner awareness. I tend to the wounds of my past and present. My journey of healing allows me to transcend my traumas. I aim to be resilient and aware. I help others who are going through their own healing journey.

Thyme – openness to love, peace and happiness

My intention is to reclaim my wilderness. I open myself up to connectivity and to experience bliss in the form of love, peace and happiness. I always remember my free will, because this allows me to be wild and free. I am in an open conversation with the universe as I practise gratitude and share my abundance with the world.

Step 2. Choose the ritual

There are different ways to go about setting an intention. Here are some ideas for rituals using herbs and flowers.

- ☮ Planting intention – plant the seeds of the plants and cultivate them.

- ☮ Vaporising intention – vaporise essential oils to clear and purify the air.

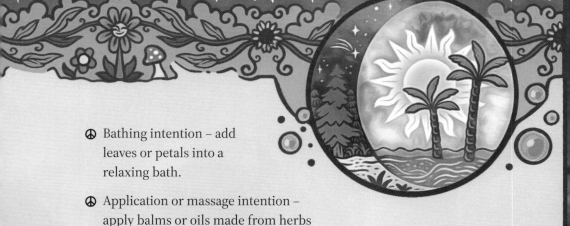

☮ Bathing intention – add leaves or petals into a relaxing bath.

☮ Application or massage intention – apply balms or oils made from herbs and flowers.

Step 3. Voice your intentions

☮ Plant both feet firmly on the ground and start the ritual (choose one above).

☮ Envision your intention, let it be a spark of energy.

☮ Bring to mind feelings of love, peace and happiness.

☮ Voice the intention and make your vow to your higher nature.

☮ Close your eyes and stay mindful to let the intention sink in.

Manifesting your intentions

What we think, we become. What we seek, seeks us. What we send out, comes back to us. Your intentions (see page 146) open you up to the process of transformation, which is necessary to stay on track for the journey of manifestation. What's always of utter importance when manifesting is that you have to imagine what you want to manifest as vividly as possible, as if it is already present in your life, otherwise your dreams, plans and goals stay wishful thinking. Having this belief is how you open your eyes to see possibilities and opportunities that are within your reach. Taking them is part of the string of actions that bring you onwards and upwards. Step by step you move beyond your limitations and reach new stages of your transformation.

Purpose: your wild pursuit

Being wild at heart and free in spirit, you have the power to choose where to be, where to go and where to stay. Only you know what you really long for. Only you can discover what your potential is and know what will make your life's journey worthwhile.

Your purpose can show itself as the thing that makes you want to get up and live life every morning, or it can be the thing that makes you feel driven. Look to your own passions – they are your fuel. To figure out what they are you can ask yourself: 'What do I burn for? What sparks my interest? What makes me feel energised?' Knowing what drives you, you can stay positive, dedicated and energetic which is important to continuously get to the places you want to go. You don't have to limit yourself to one single purpose; you can have several. But remember not to pressure yourself – it's about what comes natural to you. Your purpose should never compromise your overall wellbeing or lead you to burning out!

I would like to mention here that your priorities might change over time, and that this can affect your sense of purpose – like it did for me. First, it was travel, then creativity and self-expression, now it's moving towards community and helping others. I'm allowing my purpose to be nuanced and flexible, so that it flows with the different phases of my life.

Vow to practise openness

I follow what makes my higher nature
grow and shine, with fire, water, earth,
air and spirit I find what's mine.
I go above and beyond to see
within and without.
As I open myself to everything,
I let go of worry and doubt.

ACTUALISATIONS – FROM DREAM TO REALITY

Knowing your intentions and what drives you, you are mentally prepared to get going, but autopilot is not an option. You need to pay attention to the road and invest your time and effort to turn your dreams into reality. That's how the journey beyond can become the most epic adventure you have ever embarked on.

I have a history of turning my dreams into reality – some might think I've just been lucky, but it took a whole lot of work, skimping, saving and planning to make all my dreams come true. When it came to our travels, Lenny and I had to constantly compromise and prioritise what we did with our money and time. In the beginning, we were a little careless with money, but with time we learned how to manage money better in order to make our trips last as long as possible. We avoided costly experiences unless we really wanted to do something specific, we ate on the cheap, rarely stayed at campsites and when we backpacked in Asia, we opted for the cheapest accommodations. We always thought twice about how we were spending our money, and were willing to make compromises whenever we could.

I believe achieving goals begins with adopting the right mindset – one that is grounded in reality but optimistic by nature. Although it would be cool to hitch a wagon to a star and change your reality within the blink of an eye, this is not how dreams usually come true. No matter if you want to master a

trait, change your reality or simply want to find balance in your current life, you need to look at what opportunities are within your immediate reach and take it from there.

Coming up are a few tips that I hope will help you begin to take the right actions to set your intentions and goals into motion.

Pinpoint your goals

1 Write down the dreams, goals and projects you'd like to obtain or achieve by the same time next year.

2 Think about why you have these ambitions. In which ways do you think they are important? Will they positively affect your everyday life or the lives around you?

3 Think about how you'll achieve your goals. How much time do you need to set aside? What are the financial costs? What tools, courses or connections will help you get there?

4 Contemplate an action plan. Break down your goals into specific parts.

5 Keep a goal journal and write in it every day. Try to focus on the positive aspects of your journey towards your goals, but also write down your let-downs and set-backs.

Activity: Write down your goals

You have probably heard that a goal that's been written down is more likely to be pursued. These are the findings of Dr Gail Matthews, who did a study on goal achieving. In the sample group, people were 42 per cent more likely to achieve a goal if they wrote it down. So literally taking note of your goals will make you more likely to go through the steps to actualise it.

Virtues to make your dreams come true

Positivity

Whatever you wish and aim for may lead you through times and places of discomfort. You must prepare yourself to go through ups and downs, and for the winding roads and the possibility of detours and dead ends along the way. Focus on keeping an open mind, an open itinerary and an open schedule, because things like goals can't be rushed.

Productivity

Accomplishments are a result of productivity. You can be most productive when you focus on keeping your needs covered. Stay healthy, take breaks and find happiness in taking the journey. Your wellbeing shall not be compromised by your goals. After all, you want this process to transcend you, not drag you down and drain you of all your energy and happiness.

Persistence

Reaching your end goal might go beyond your current capabilities, but don't let that discourage you. When you begin to aim towards something, you start a process that changes, evolves and transcends you in ways you could never have imagined. Whenever you have a dream or goal, always factor in that you can pick up abilities, lessons and virtues along the way that will allow you to get where you need to go.

Patience

Aiming for a goal starts with taking one step and then the next. Step by step you can get closer to reaching your end goal, and you may need to take a thousand steps or more. It might take double or triple the time or effort you initially expected, but if you really believe in your goal, it will be worth all the time and hassle. Visualise yourself as having virtues in the form of resilience and dedication to find patience within.

Keep it real – the grass is not greener

We are sold dreams left, right and centre and all day long through magazines, TV and social media. Ideas of ideal bodies, ideal relationships and ideal lifestyles make us believe in the idea of an ideal life. While we occupy ourselves with believing in the idea of a perfect life, we are forgetting we are basing this on merely snippets and snapshots of other people's seemingly fabulous lives. Our vivid imagination fills the gaps and so we believe a beautiful fantasy that has no merit in the real world.

Craving perfection from the inside out, we might think we need to find the perfect match, create a perfect family and live in a perfect home with a perfect view to be happy and feel complete. Striving towards perfection always creates new reasons to feel incomplete, because after all there is no such thing as perfection. It is an out-of-touch idea created by the imagination of our resentful and miserable egos. Comparison often leads us to think the grass is greener somewhere else, but if we would just pay a little more attention to ourselves, we would see that the grass is green where we water it.

I learned this lesson through the wonder and disarray that is van life, which has become famous for its posed photos and romanticised narrative. Even though we'd been on bigger trips with our old van, Elefriend, and knew the stress it entailed in the long run, the pretty photos sold us the idea that what we needed was just a bigger van that would allow us to live in the van full-time. One year and an expensive van-build later, I started to realise that no matter how much I loved the idea of living the van life, it just wasn't a full-time option for me, and I was actually more stressed and had less free time than before. Since we moved into an apartment, I live a more hassle-free life, and get to focus on watering my own lawn, so to speak, and when we go on trips with our van, I get to enjoy it more fully, as I no longer try to live up to perfection.

Be content while ambitious

No matter what you want to achieve or do, remember that it's in human nature to be ambitious, but being on a constant chase to achieve or obtain something won't lead to happiness, because it means that you never learn to be completely satisfied with the circumstances. That's why it's important to learn to be content in your current reality, no matter if you are staying put or are on an ambitious journey. Whatever your goals are now or in the future, I wish you the best of luck!

Daily reminders to help align the stars

Celebrate every achievement, no matter how small.

A little progress is still progress.

What isn't started today cannot be finished tomorrow.

Aim to be happy and healthy, no matter where you are.

ELEMENTAL MOMENTS

The most important moment of life is always the moment of now. This is the only time that you can live and breathe, think and feel, act and react. Living mindfully in the now does not mean disregarding all that lies in the past. Memories are a vital part of our journey in life, and when we look back on them we can reminisce about a certain moment in time or gain a deeper understanding of ourselves, giving us more clarity of who we are now. Thinking back can especially help in times of stress and disarray, because we can bring our thoughts to good times, which can motivate us to pursue more good moments. Just like every other moment has come and gone, we can take comfort in the fact that everything is temporary, and by practising a little faith, we can calm our nerves in our darkest moments, knowing brighter days are usually ahead.

Thinking about our memories, we can also contemplate on how far we've come and see where and how we made progress on our journey. They can serve as long-lasting bridges to feelings and experiences that we've had in the past.

Activity: Capture memories

If you want to remember what you felt and experienced at a particular time, your best bet is not to wait too long before capturing it. The more time passes and the further your mind has to go back, the more likely it is the memory will change, so capture the moment when it's recent and vivid. Choose a photo, it can either be your own or one you think represents the moment well, and print it to turn it into a tangible object. Write down your memory on the back of the photo, and include the five Ws. Who is it about? What happened? When did it take place? Where did it take place? Why did it happen? And lastly, add How it made you feel.

Marking moments

Particular moments can be made significant when an occasion is marked or celebrated, and this practice of celebrating a point in time has been passed down to you in the form of rites and traditions. Many cultures make a big point out of certain life events, like birthdays, weddings and funerals. Often, we only celebrate the occasions that have been passed down to us, but there's no monopoly on ceremony and it's not reserved to what is commonly celebrated, so look into what other moments you think should be marked. In this way, you can make a point out of anything you want to give more importance or significance to.

It's simple, you turn a moment into an occasion to mark its significance. Do so when you reach milestones, go through healing journeys and go through rites of passage. You can either create a personal ceremony or plan one together with others. It can be as short-lived as clinking glasses together or be a week-long festival. By marking an occasion, you can honour all kinds of moments, good and bad, happy or sad and everything in between.

167

How to mark moments

When you begin the ceremony, you can voice the purpose of this ceremony: 'I commence ... to mark ... in order to ...'. It's up to you to fill in the blanks. You can then carry on with your commemoration by singing, dancing, feasting on food, sitting in quiet contemplation, meditating or writing down your hopes and dreams. You are wild and free to do whatever feels right to mark a moment, so use your wonderfully wild side (creativity, imagination and playfulness, see page 94) as part of the planning process.

Ideas for moments to mark

☮ Reaching a milestone.

☮ Overcoming an obstacle.

☮ Reaching a goal.

☮ Entering a new life phase.

☮ Beginning and concluding a trip.

☮ Going through recovery.

☮ Healing of an injury or illness.

☮ Seasonal events, like solstice or equinox.

☮ Lunar phases, such as full moon, new moon or eclipse.

Elemental ceremonies

Earth
Decorate with greenery, flowers, branches, stones and gems.

Fire
Light candles and incense, place lanterns and create a warm atmosphere.

Air
Air out the room or stay outside, and open ceremony with a song, chant, mantra, affirmation or ringing a bell.

Water
Stay rejuvenated by drinking water, and wash yourself before you mark the moment.

Spirit
Be present by staying mindful and intentional for the duration of your ceremony.

ELEMENTAL CONCLUSIONS

Our life's journey is constantly in motion and nothing ever stays the same. We go through life's seasons and grow at our own pace as we evolve through time and space. On one level or another, we are always changing, yet our mind mainly registers immediate change and visible shifts in situations. This means we might sometimes miss paying attention to parts of our transformation or can't make sense of the constant changes, making us feel lost or adrift within our jungle of a mind.

After my first couple of years on the move, I had already been to so many places, met so many people and done so many things that my mind became a blur of all kinds of impressions. That's why I started to think about how I could find a way to round up my experiences to have closure about things that happened in the past and to gain more mental clarity in the present.

I now make a habit of going through the different stages of reap, reflect, remind, recharge and revel. The roles of each stage can be simply explained by using the elemental forces of earth, water, air, fire and spirit (see page 58). Combined, they allow me to stop and look, evaluate and, lastly, move on. That's how I make the most of every step of my growth process. I like to use these lessons every day, but I find them to be especially helpful when there are bigger shifts happening to make note of what's really going on.

Earth – Reap

Overcoming an obstacle. Completing a task. Finding a solution. Realising a goal. Reaching a milestone.

The element of earth teaches us to be sustainable and to reap what we sow. The moment we obtain what we've been aiming for and our plans come into fruition is the moment we can reap what we've manifested. Working towards this point and arriving here, we shall enjoy the riches, the abundance and the bountiful rewards. Enjoy the taste of success and savour it. Don't devour it or gobble it up, as it'll only make you look for the next thing to dine on – truly appreciate what you've reaped in this very moment.

Water — Reflect

Contemplate the past. Learn from mistakes. Find inner wisdom. Grow from the lessons. Make improvements. Relive memories.

The element of water teaches us to reflect on our experiences and dive into our emotions. Take the time to look back to see how far you've come on any given journey. Think about the detours, the bumps and dead ends. Ponder on everything that's happened and bring to mind all the good, the bad and everything in between. Try your best to make sense of your journey and reflect on everything it entailed. What went right or wrong? Is there something you would have done differently? Are there things you should research to gain more clarity? This is how you can learn your lessons and make improvements for all the journeys to come.

Air – Remind

Note findings for the future. Keep track of your process. Record your journey. Memorise your intentions. Connect past, present and future.

The element of air teaches us to use our inner findings from reflection and to contemplate what we want to bring with us into the future and what we want to leave in the past. Think about how past lessons can be utilised in ways at another time and place. Is there something you want to remind yourself of so you don't forget it and repeat a mistake? Use your notepad and jot down what you want to remind yourself of going forward. Write down what you want to do differently. This will help steer you in the right direction on future journeys.

Fire – Recharge

**Rest and rejuvenate. Kick back and relax.
Take it easy. Escape from stress and to-do lists.
Energise yourself.**

The element of fire teaches us to tend to our flame. You have
come this far, you reached your goal, reflected on the process
and learned your lessons. Now you might have a new goal in
sight, but before you continue onto the next leg of your journey
you need to recharge. Being in constant motion, we are in
constant transformation. Nothing ever stays the same, and
we need time to adjust to all the changes that are happening
to cope with all the things we go through. We are constantly
fed with impressions and these need time to be digested,
otherwise we stress ourselves out and can eventually burn out.
Set aside time to take a breather to recharge your batteries;
that's how you fuel up and get ready for a new string of actions.

Spirit – Revel

Celebrate where you are at. Find joy in taking the journey. Honour the process. Count your blessings.

The element of spirit teaches us that no matter where we are on our journey, we should appreciate where we're at and how far we've come. It's often difficult when we have our eyes set on something that we think would complete us in the future. This is when you need to find it in yourself to see that the journey is the destination. Here, you learn and live. Don't keep asking yourself, 'Am I there yet?' Instead, reassure yourself: 'I am where I'm supposed to be.' You are always in transformation, and no matter what phase you are in, there's love and light, growth and delight – even in times of hardship and difficulty. Always look at the bright side of life to uplift yourself and revel in each and every moment.

THE NATURE OF ALL

You are the centre of your existence, and what surrounds you is everything and everyone else. You might only be one person, one being, one earthling, but you are a part of a bigger picture. It's time to zoom out of yourself and look at the masterpiece of life. Open your eyes and you will see yourself in the natural work of art called the universe. A vast place that is always expanding, changing and growing – just like you. It's darkness and light, it's despair and delight, it's love and ever so bright.

Yes, you are one of a kind, one of everyone and one of the wholeness that is the universe, and your existence is interwoven into the grand tapestry of life. That tapestry is not all sunshine and rainbows, sometimes it's the complete opposite. I bet you've also experienced dark moments or phases that made you feel disconnected from yourself, from others, from everything.

Just as connectivity is natural, so is the feeling of disconnection. After all, we would not know one if we didn't know the other. I know disconnection all too well, and have felt it time and time again, whether it was because of my dysfunctional family, my egotistical pursuits going nowhere or the moments I lacked self-love. The only thing that got me out of the darkness was opening myself up to the light and asking myself, 'Am I really alone? Am I unloved? How am I disconnected?'

Whenever you feel disconnected remember this: together with billions of other humans and trillions of earthlings and gazillions of living organisms, you get to take an epic ride on Earth on a gorgeous road called the Milky Way. So while you

are here, aim to find connections and cultivate these, because what makes life ever so precious are the beautiful beings we take this wild journey with!

Channel the nature of all

You can onset connectivity and transformation by channelling the nature of all. Coming up are three different notions of wholeness followed by a call for oneness. When you feel disconnected, lonely or scattered, you can read this out loud. As you voice this, imagine how you unify yourself with your community and with the universe.

Oneness: the universe

Oneness, also called the spirit, is the essence that connects us to everything. When living in oneness, we can understand that we are not separate individuals. All together we are one. This is the nature of all. It goes further than human ranks, and includes all other earthlings, all of nature, all the way above and beyond, connecting ourselves to everything in the universe.

Everyone: the collective

Oneness is what creates unity in many constellations, and it allows us to create connections, from family and friendships to communities and collectives. Since we are deeply dependent

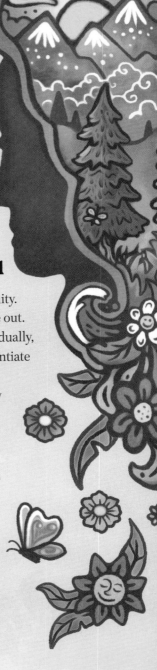

on each other, oneness through a collective spirit is how we can best work towards common goals for the greater good. A happy coexistence with other individuals is how we can form relationships that help us survive and thrive – that is the power of togetherness and a collective spirit.

One of a kind: the individual

We are one of all, one of the human race, one of a community. On an individual scale, each of us is unique from the inside out. Our bodies are similar but not completely alike, and individually, we possess a unique set of traits and memories that differentiate us from others. Combine that with our talents, interests and ideas, and we become one of a kind – an extraordinary manifestation created by the nature of all.

A call for oneness

I open myself to my wilderness,
and I embrace my higher nature.
I live in wholeness with others,
because together we are greater.
I share my love, hope and peace,
as I revel with joy and happiness.
Here and now, I become one with all,
unified by the words that I've called.

One for all and all for one

I've watched the sunrise over an active volcano, dived in colourful coral worlds, cycled amidst ancient temples, and boated offshore epic coasts. I've driven roads that were terrifying, walked paths that were unsafe and spent many nights amidst the elements of nature. I experienced moments of bliss witnessing the beauty of life, and lived through moments of despair that made me stronger and more resilient.

Even after all that I have been through, my most treasured memories are of the beautiful and wild people I've met along my journey. Connecting with people has made for some of my happiest and brightest moments in my life. Making friends from all around the world made me realise how much we all have in common, despite our different backgrounds, cultures and upbringing. Despite our differences we could still share love and light, and laugh at the weirdness of life.

I think the fact that we have more in common than what separates us is something worth remembering by each and every one of us, as we are living through very divisive times. We need to think beyond ourselves and look at the bigger picture, so we can work towards justice, equality and equity. We need to work together in the direction of a common ground; that is how we create peace and harmony on a collective scale. Now, I am obviously an idealist, but you might also count yourself a dreamer and believe a better world is possible. If yes, then you

We are better together

Always and forever

can use your good intentions and be the change that you wish to see in the world.

Show up for your fellow humans – help, support and uplift them. Listen to their needs and try to connect with them on a deeper level. While connecting focus on common ground, celebrate inclusivity, cooperate and aim for togetherness. Cultivating human relationships can give you the courage and strength to see what other good things you can do for humanity, earthlings and nature as a whole.

Natural order

This whole book, I've been encouraging you to think beyond your ego and connect with your wild and free spirit. I motivated you to get out into nature and strengthen your bond with the natural world, as reconnecting to the nature within and around you is a means to restore your own natural order and gain back what was either lost or forgotten. As you live in your wholeness, unified with your body and spirit, you can now look at the bigger picture and see how you can help restore natural order on a grander scale.

When looking at the current state of the world, it is evident that its natural order is becoming a mess, due to human intervention. Reclaiming your natural self, you can use your new-found connection, strength and awareness to give back to mother nature – after all, she's the reason you're here. It's no secret that nature's an important part of our lives.

PLANET EARTH

OUR HOME THAT WE MUST
PROTECT AND LOVE

In fact, we are and always will be completely dependent on it. Being interwoven into the nature of all things, we rely on nature's ecosystems, and any imbalance or destruction can very well lead to devastating consequences, which they sadly already are.

The vulnerabilities of the natural world are very visible in this day and age, with climate change, species extinction and the constant loss of natural habitat. These are just some of the signs that the natural systems are in disorder. Humans are harvesting natural elements, calling them natural resources. We are genetically modifying animal species in the name of supply and demand, and we continuously acquire natural territories for the sake of human purpose. But nothing in nature belongs to us.

On the contrary, we belong to nature, and when we accept this, we can make better decisions for ourselves and our planet. If we begin to see ourselves as the stewards of nature, instead of as owners, we can take actions that are in the best interest for all. This is how we can choose evolution above self-destruction.

Even if you're not the one that is chopping down the rainforest or contaminating waters with chemical waste, your choices might pay an industry that does just that. Every dollar you spend has an impact, so choose wisely to either minimise or maximise the impact of your purchases. By combining our efforts, we can work towards a world of non-violence, and the sum of our actions can help serve and protect planet Earth and all its inhabitants.

Do your part

Your everyday actions affect the world and have an impact on other people, animals and nature. Nature needs your helping hand so that it has a chance at being preserved and protected. Fellow humans need your consideration, so uplift them rather than disadvantage them. Animals need your empathy, so don't exploit them for entertainment, clothing or food. It's virtually impossible to not cause any damage at all, but this shouldn't stop you from making conscious choices every day and aim to improve your daily habits, including what you buy, what you wear, what you eat, the trash you produce and the transport you use to get around. All of what you do and don't do affects the world. Your time, consideration and extra effort will benefit all those around you, your local environment and the whole world. This is a call for change, and each positive change you do will help secure future generations.

DO YOUR PART

Be considerate

☮ Reduce, reuse and recycle your waste – especially plastics.

☮ Fight food waste by saving leftovers for later and using food-sharing apps.

☮ Take part in a beach clean-up to gather trash and keep nature clean.

☮ Avoid flying, take public transport and carpool when possible.

☮ Be an eco-tourist by supporting local businesses and respecting the local environment.

Be responsible

☮ Rent or share items you rarely need, like vehicles, tools and adventure gear.

☮ Buy used instead of new from flea markets, second-hand shops and online marketplaces.

☮ Be conscious when purchasing – buy quality over quantity.

☮ Go plant-based to cut your environmental footprint in half.

☮ Buy local and seasonal produce, and buy organic when possible.

Be compassionate

☮ Show up for your fellow earthlings and help uplift them.

☮ Learn about your own local history and world history.

☮ Challenge your perspective, your beliefs, your judgements and your ideas.

☮ Widen your circle of compassion to include all humans and living beings.

☮ Don't attend animal shows, zoos or other activities that exploit or disturb animals.

GOING FULL CIRCLE

Nature is both chaos and order. It's a universal law, making the world go round, both your inner world and the world as a whole, and mentally, physically and metaphysically speaking. Chaos and order are the natural way of things, and it's the natural way of life. It's the fine balance in between that creates harmony. When our life is too chaotic, stressful and challenging, we are out of balance, but the same goes if our life is too easy, comfortable and carefree. In fact, everything has its opposite, everything is dual and we cannot accept something entirely without accepting its equal opposite.

Yin and Yang

The philosophies of Taoism teach us that every aspect of life is created from a balanced interaction of opposite and complementary forces. They call this Yin and Yang, and together they create wholeness, as they transform each other and their balanced interaction gives birth to new things.

Even though Yin and Yang are competing, they do not cancel each other out. Every rise transforms into a fall and every advance transforms into retreat. We wouldn't be able to understand one energy if we didn't know its opposite, like cold and warm, day and night, giving and receiving. Taking the example of dark and light, they both have their individual nature, but you can't understand this until you know their opposites – you can't really know what one is without the other. But what does this all mean for you? Simply speaking, accept the bad with the good and see the good in the bad, but a deeper take away is to always aim to create harmony within.

YIN — YANG

Yin	Yang
night	day
cold	hot
earth	sky
moon	sun
slow	fast
calm	expressive
dark	light
internal	external
receptive	creative
death	birth
quiet	loud
order	chaos
asleep	awake
stillness	movement
autumn/winter	spring/summer

Your balanced wholeness

Attempting to live in your wholeness is how you can work towards balance in mind, body and spirit, but know that you'll never reach a point of everlasting balance. Since you are in constant transformation and always exposed to outside circumstances that affect your life, you continuously need to recreate this balance as you grow and evolve. You can attempt to live in balanced wholeness, but it'll always need work, attention to detail and full awareness. I encourage you to keep pursuing your wild and free nature, because being true to yourself is how you can create balance and find bliss again and again.

All endings are also beginnings

Millions of years of evolution has led to this point in existence. Everything that has happened to you and every choice you have made along your life's journey has shaped you into the beautiful person that you are. If there's one thing that I hope you take away from this book, it's that you should always try to stay open to new perspectives, new opportunities and new lessons, as you never know what wonders or wisdoms you might find. I hope my suggestions and findings have helped you find ways to explore your own nature and nature as a whole. As you journey onwards, I hope you will enjoy living on the wild side and that you keep connecting with the spirit of nature. Just remember: no matter where you go and what you do, don't stop dancing to the beat of your own drum! Your wild spirit shall revel and be free, because that's how you are truly meant to be!

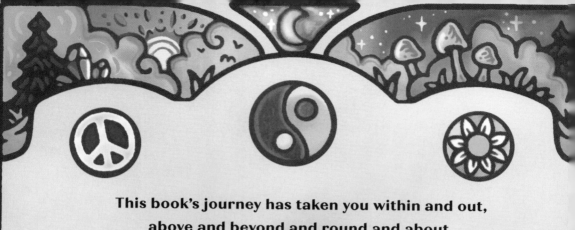

This book's journey has taken you within and out,
above and beyond and round and about.
Your pursuit for wilderness will never be done,
heck, it might have only just begun!

Keep yourself open and look for inspiration,
be your true self as you seek exploration.
Set your intentions and strive for your goal,
unleash your spirit in body, mind and soul.

Stay mindful and kind and pave your own way,
appreciate life's wonders each and every day.
This is my last reminder for you to see,
it's in your nature to be wild and free.

I now wish you good luck on your onward adventures,
may your journey always be filled with love
and splendours.

And on that note, I wave goodbye.
See you in another place, in another time!

Resources

The nature of you

Eckart Tolle. *A New Earth*. 2005, reissued 2018. New York, N.Y. USA.
Dutton/Penguin Group // The Current State of Humanity 25–34

Choose your own path

Eckart Tolle. *A New Earth*. 2005, reissued 2018. New York,
N.Y. USA. Dutton/Penguin Group // The Core of Ego 59–82,
Role-playing: the Many Faces of the Ego 85–94

Navigating through life

Medical News Today. *What to know about amygdala hijack*. 2022.
medicalnewstoday.com/articles/amygdala-hijack#what-is-it

Live mindfully

The National Center for Complementary and Integrative Health
(NCCIH). *Meditation and Mindfulness: What you need to know.*

2022. nccih.nih.gov/health/meditation-and-mindfulness-what-you-need-to-know

Zen Buddhist Abbot Shohaku Okumura – Zazen is good for nothing. 2018. youtube.com/watch?v=8T-Z1WoFXkk

Unearth your nature

Eckart Tolle. *A New Earth*. 2005, reissued 2018. New York, N.Y. USA. Dutton/Penguin Group // The Pain-Body 131–157

Ian Morgan Cron and Suzanne Stable. *The Road back to You*. 2016. Brentwood, TN USA. InterVarsity Press

Be in your element

Harvard Health Publishing. *How much water you should Drink*. 2022. health.harvard.edu/staying-healthy/how-much-water-should-you-drink

Dr. Michael Greger with Gene Stone. *How not to Die*. 2015. New York, N.Y. USA. Flatiron Books

World Health Organisation. *Physical Activity*. 2022. who.int/news-room/fact-sheets/detail/physical-activity

Environmental Working Group. *Reduce Toxins*. 2017. ewg.org/news-insights/news/5-ways-reduce-toxic-exposures-your-home

The five elements

Sarah Durn. *The Beginners Guide to Alchemy*. 2020. Emeryville, CA USA. Rockridge Press // Alchemy and the Elements: 25–32

Brian Cotnoir. *Practical Alchemy.* 2006 (under the name The Weiser Concise Guide to Alchemy by Red Wheel Books), reissued 2021. Newburyport, Ma. Weiser Books // The Aspects of Alchemy 3–9

Glennie Kindred. *Elements of Change.* 2009. Derbyshire, GB. Self-published

Joseph M. Seguel. *Indoor Air Quality.* 2017. ncbi.nlm.nih.gov/pmc/articles/PMC6125109/

Nursery Live. *Top 10 Highest Oxygen Producing Indoor Plants.* 2022. nurserylive.com/blogs/top-10-plants/top-10-highest-oxygen-producing-indoor-plants

Krishna Savani and Satishchandra Kumar. *Beliefs about Emotional Residue.* 2011. pubmed.ncbi.nlm.nih.gov/21688925/

It's about time

Healthline. *How Long Does It Take for a New Behavior to Become Automatic?* 2019. healthline.com/health/how-long-does-it-take-to-form-a-habit

Reconnect with nature

Healthline. *Spending Time In Nature Is Good for You.* 2022. healthline.com/health-news/spending-time-in-nature-is-good-for-you-new-research-explains-why

Mind. *How can nature benefit my mental health?* Retrieved 2022. mind.org.uk/information-support/tips-for-everyday-living/nature-and-mental-health/how-nature-benefits-mental-health/

Go camping

LNT. *Leave no trace principles*. Retrieved 2022. lnt.org/why/7-principles/

Invite nature in

NRDC. *Bringing the Outdoors In: The Benefits of Biophilia*. 2020. nrdc.org/experts/maria-mccain/bringing-outdoors-benefits-biophilia

Frank H. Mahnke *Color, Environment, and Human Response*. 1996. New York, N.Y. USA. John Wiley and Sons, Inc

Live intentionally

National Library of Medicine. *A history of chaos theory*. 2007. ncbi.nlm.nih.gov/pmc/articles/PMC3202497/

Veritasium. *Chaos: The Science of the Butterfly Effect*. 2019. youtube.com/watch?v=fDek6cYijxI

Actualisations – from dream to reality

Dr. Gail Matthews. *Goals Research Summary*. 2015. dominican.edu/sites/default/files/2020-02/gailmatthews-harvard-goals-researchsummary.pdf

Going full circle

BBC. *Concepts within Taoism*. 2009. bbc.co.uk/religion/religions/taoism/beliefs/concepts.shtml

John Bellaimey – TED-Ed. *The hidden meanings of yin and yang*. 2013. youtube.com/watch?v=ezmR9Attpyc

Acknowledgements

This book's journey began ten years ago, when I set off on my first adventure – everything that has happened and everyone who I've met since then has shaped me and my story. I would like to share my thanks to all those who were part of this journey, and who contributed in their very own way.

First and foremost, I would like to thank Lenny – my partner, my love, my everything. Every step of this wild adventure was taken together with you, and I look forward to continuing to break away with you as we go here, there and everywhere.

Next, I would like to thank the Hardie Grant Explore team, especially Melissa and Amanda. You have been patient, supportive and encouraged me to tell my story. I also want to thank you both for connecting me with Helena, my amazing editor and fellow Dane and adventure enthusiast. Working with you was a breeze and you helped me bring my vision and ideas to life.